The Science on Women and Science

The Science on Women and Science

Christina Hoff Sommers
Editor

The AEI Press

Publisher for the American Enterprise Institute

WASHINGTON, D.C.

Distributed to the Trade by National Book Network, 15200 NBN Way, Blue Ridge Summit, PA 17214. To order call toll free 1-800-462-6420 or 1-717-794-3800. For all other inquiries please contact the AEI Press, 1150 Seventeenth Street, N.W., Washington, D.C. 20036 or call 1-800-862-5801.

Library of Congress Cataloging-in-Publication Data

The science on women and science / Christina Hoff Sommers, editor.
    p. cm.
  Includes bibliographical references.
   ISBN-13: 978-0-8447-4281-6
   ISBN-10: 0-8447-4281-3
   1. Women in science. 2. Sex discrimination against women. 3. Sex differences in education. I. Sommers, Christina Hoff.
  Q130.S364 2009

                                    2009022004

13 12 11 10 09   1 2 3 4 5

Printed in the United States of America

Contents

List of Illustrations

Figures

Tables

Introduction
The Science on Women in Science

Christina Hoff Sommers

Are women victims of a widespread bias in science and engineering, as a 2007 report of the National Academy of Sciences (NAS) concluded?[1] Or can the paucity of women in various quantitative fields be otherwise explained? What, if anything, should be done to encourage more women to become engineers and scientists?

In the fall of 2007, the American Enterprise Institute brought together several outstanding scholars to discuss and debate the controversies surrounding the topic of gender and science. We also commissioned a paper by psychologists Jerre Levy (emerita, University of Chicago) and Doreen Kimura (Simon Fraser University) to react directly to some of the key findings of the NAS report, *Beyond Bias and Barriers: Fulfilling the Potential of Women in Academic Science and Engineering*. Scholars, legislators, journalists, and others looking for a balanced and temperate treatment of this sometimes contentious topic will welcome this collection of the papers prepared for the AEI conference.

Some women's advocates warn that debates over sex differences are dangerous for women. Such discussions, they say, serve only to reinforce discredited stereotypes. When *ABC News* aired an hour-long special in 1995 highlighting new research on sex differences, prominent feminists such as Gloria Steinem and Gloria Allred were strongly critical.[2] Why, they asked, was the network giving currency to damaging and discredited research? According to Steinem, such research "is what is keeping [women] down." Allred warned the ABC interviewer, "This is harmful and dangerous to our daughters' lives, to our mothers' lives."[3]

A decade later, Lawrence Summers, then president of Harvard University, met the same passionate disapproval when he referred to a body of research that suggested men and women might not have identical interests and propensities. A prominent member of the psychology department told the *Harvard Crimson*, "In this day and age to believe that men and women differ in their basic competence for math and science is as insidious as believing that some people are better suited to be slaves than masters."[4]

But beliefs about sex differences in math and science ability cannot usefully or seriously be compared to beliefs about slavery. No dire consequences follow if it turns out that women—as a group—are more adept at, say, teaching, pediatrics, law, and veterinary medicine than at engineering, computer programming, and automobile repair. The literature on gender and vocation is complex, vibrant, and full of reasonable disagreements, and it encompasses differences in interests and career preferences as well as in innate abilities. There are sensible and fair-minded scientists on all sides. They should be free to argue without being intimidated, silenced, or compared to racists.

In his classic essay, *On Liberty*, John Stuart Mill wrote about the hazards of silencing dissident opinions: "If the opinion is right, [people] are deprived of the opportunity of exchanging error for truth; if wrong, they lose, what is almost as great a benefit, the clearer perception and livelier impression of truth produced by its collision with error."[5] As readers will see, the debate over women and science is still very much unresolved, and we cannot yet separate error from truth.

It is regrettable that, without the benefit of the kind of free and open debate Mill was advocating, we find the U.S. Congress and federal agencies such as the National Science Foundation (NSF) and the National Aeronautics Space Agency (NASA) moving ahead with ambitious remedies to deal with the putative bias against women in the academic sciences. On October 17, 2007, a subcommittee of the U.S. House Committee on Science and Technology convened to learn why women are "underrepresented" in academic professorships of science and engineering and to consider what the federal government should do about it.[6] All of the expert witnesses and every member of the committee, Republican and Democrat, were in complete accord that sexist bias was the reason there are relatively few women in university science. Nor was there much dispute about the remedy. The committee strongly supported the solutions proposed by the National Academy of Sciences' *Beyond Bias* study. It recommended workshops to educate federal and academic personnel about unconscious bias and ways to combat it, and suggested revisions to universities' criteria for evaluating academic advancement. It also urged federal agencies to conduct stringent Title IX compliance reviews of math, science, and engineering programs. Those reviews are already underway.[7]

Some of the authors included in this volume agree with the NAS study and applaud Congress for its efforts to counter sex bias in academic science. They point to studies that find little or no difference in the core cognitive abilities of men and women; they reveal how exposure to sexist stereotypes can undermine women's performance; and they highlight the powerful human tendency to divide the social world by gender—often unjustifiably treating one sex as superior to the other. But other scholars believe that the NAS efforts are premature and may do more harm than good. These authors are not convinced that the scarcity of women in the hard sciences is due to ongoing discrimination; they attribute the gender disparity to characteristic gender preferences grounded in biological differences.

In arranging the conference and inviting the essays for this collection, my AEI colleagues and I sought to find the best proponents of the various positions in the controversy. Our hope is that the conference and the essays will inspire other educational institutions to encourage open discussion of the causes of the numerical disparity. Members of Congress might wish to convene hearings and commission studies in which both sides of the debate are represented. Before initiating aggressive Title IX reviews of physics, math, and engineering departments, and before providing more funding for programs to eliminate bias, we should be sure that bias is the problem.

As a philosophy professor and equity feminist in the classical liberal tradition, I am well aware of the long and sorry history of how alleged natural differences between men and women have been routinely and casually interpreted by men as proofs of their superiority to women. Often the claims of difference were absurd; but almost always, women paid a heavy price. It is understandable that today many women and men, keenly aware of that history, continue to react with suspicion to the suggestion that the sexes are in any significant way innately different. Nevertheless, the corrective to the history of damaging bias is not more bad science; it is good science, clear thinking, and open, fair-minded discussion. Hence, this collection.

Notes

1. National Academy of Sciences et al. 2007.
2. *ABC News* 1995.
3. Ibid.
4. Hemel 2005.
5. Mill 1978, 16.
6. U.S. House of Representatives 2007.
7. See Tierney 2008 and Sommers 2008.

References

ABC News. 1995. Men, Women, and the Sex Difference. *ABC News Special.* February.

Hemel, Daniel J. 2005. Faust to Lead New Initiative. *Harvard Crimson.* January 24. http://www.thecrimson.com/article.aspx?ref=505409 (accessed January 12, 2009).

Mill, John Stuart. 1978. *On Liberty.* Indianapolis: Hackett.

National Academy of Sciences, National Academy of Engineering, and Institute of Medicine of the National Academies. 2007. *Beyond Bias and Barriers: Fulfilling the Potential of Women in Academic Science and Engineering.* Washington, D.C.: National Academies Press.

Sommers, Christina Hoff. 2008. The Case Against Title-Nining the Sciences. *Teachers College Record.* Columbia University. July 15. http://www.aei.org/publications/filter.all,pubID.28694/pub_detail.asp (accessed January 12, 2009).

Tierney, John. 2008. A New Frontier for Title IX: Science. *New York Times.* July 15.

U.S. House of Representatives. Committee on Science and Technology. Subcommittee on Research and Science Education. 2007. *Hearing Charter: Women in Academic Science and Engineering.* http://democrats.science.house.gov/Media/File/Commdocs/hearings/2007/research/17oct/hearing_charter.pdf (accessed January 13, 2009).

1

Why So Few Women in Math and Science?

Simon Baron-Cohen

It should go without saying that, along with most scientists I know, I would like to see equal representation of women in all areas of employment, including science and math. It distressed me greatly when I first became a Fellow of Trinity College, Cambridge, known for its long tradition in math and science, that of the two hundred or so fellows, only three were women. Like many people I assumed that this lack of equality—which still distresses me—had arisen as the result of some subtle form of discrimination or deterrent. The most common sociocultural explanations put forward for this outcome were some form of misogyny; a lack of same-sex role models for female applicants; and insufficient support during key stages of career development for women (especially with respect to pregnancy and childrearing).

Having been in this environment for over a decade, I am persuaded that any misogyny that may have existed is not currently evident, since the math and science professors I have met are liberal and fair-minded. The absence of same-sex role models remains a problem. In the math lectures, the sex ratio is at least three to one (male to female); it must certainly feel strange to be a female student in the minority, with the teachers also nearly all male. Similar sex ratios among math students are seen in most universities. While this might deter some women from joining these professions, however, it cannot be the whole story; a sex difference is seen in math scores in high school in the United States, long before such role-model factors at the university level

The author was supported by the MRC and the NLM Family Foundation during the period of this work.

have had a chance to operate. Figure 1-1 shows, for example, the average scores on the SAT math test, year by year, from 1972 to 1997. Despite annual fluctuations, males outperform females consistently.

FIGURE 1-1
SAT-MATH TEST RESULTS 1972–97

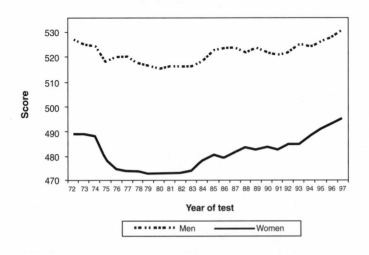

SOURCE: Figure provided by Professor Steve Pinker with kind permission.

Finally, regarding the third of these sociocultural factors, the role of support around pregnancy and child care is much improved. In academia, the job is not a nine-to-five regular office job, but typically offers flexible hours. More fathers are involved in caring for their children, and parental leave following the birth of a child is funded not just for women, but for men, as well. In addition, more fellowships have been created just for female applicants. So, without denying a long history of discrimination against women, we can say that many of these sociocultural factors are lessening in importance in today's academic world. And yet, at higher levels in universities, the ratio of men to women in math continues to be around three males to every female. Why?

For me, the clearest clue regarding this sex ratio is the roster of winners of the Fields Medal, which is often referred to as "the Nobel Prize of mathematics" and awarded to the most outstanding mathematicians under forty years of age. There has never been a female winner, despite this prize's having been awarded regularly since 1950. This fact has prompted me to ask, what is going on at the extremes of the distribution of ability in math and science? To end up with a sex ratio of one to zero among Fields medalists, either the sociocultural factors are operating even more strongly in extreme groups, or we need also to consider some nonsocial factors. To my mind, these nonsocial factors include what we could call (for shorthand) personality type and biology. I will discuss each in turn and argue that they are not mutually exclusive. A certain personality type (namely, one that is more strongly drawn to "systemize") may, for partly biological reasons, be more common in males.

In making these arguments, I will be referring to *average* differences that are found in a small way when comparing males and females in the general population. And I will also refer to a statistical property of the normal distribution that has massive effects at its extremes.[1] Renowned Harvard psychologist Steve Pinker reminds us of a surprising mathematical property of the normal distribution (shown in figure 1-2): If two groups (such as males and females) differ a bit at the center of the range (in their

FIGURE 1-2

TWO GROUPS (SUCH AS MALES AND FEMALES)
DIFFER IN THEIR AVERAGE SCORES

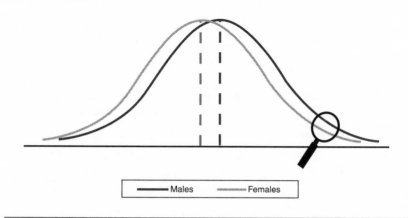

SOURCE: Figure provided by Professor Steve Pinker with kind permission.

means), then, because of the rate at which the slope of the curve falls off, the differences between them will be huge at the extremes. So, with height, for example, the two sexes differ by three inches on average. At five feet ten inches, the sex ratio is thirty to one (male to female). In people just two inches taller (six feet), the ratio jumps up to two thousand to one!

We can see quite why this is happening in figure 1-3, which blows up the portion of the distribution's right-hand tail that is indicated by the magnifying glass in figure 1-2. It becomes apparent that the gap between the sexes widens as we move to the extremes. This is a purely statistical property: The rate at which the slope falls off is a negative exponential of the square of the distance from the mean.

FIGURE 1-3

**AT THE EXTREMES, THE TWO GROUPS
(E.G., MALES AND FEMALES) DIVERGE MUCH MORE**

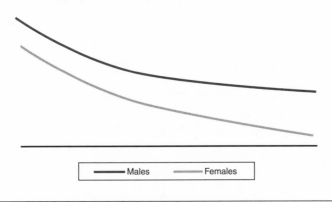

SOURCE: Figure provided by Professor Steve Pinker with kind permission.

Since the statistical rule applies to any continuous dimension that is normally distributed, it will apply as much to psychological or personality traits as to height or blood pressure. Which psychological traits might be relevant, we may ask, when we think of typical sex differences in the population relevant to aptitude in science and math? And could a small sex difference in the center of the distributions become much bigger at the extremes? Finally, could these sex differences exist for partly biological reasons?

Sex Differences in the General Population

There are interesting differences between the *average* male and female mind. In using the word "average," I am, from the outset, recognizing that such differences may have little to say about individuals. As we will see, the data actually require us to look at each individual on his or her own merits, as individuals may or may not be typical for their sex. The two relevant psychological processes in which we observe sex differences on average are *systemizing* and *empathizing*. Empathy is less relevant to the question about sex ratios in math and science, and is reviewed later. "Systemizing" is the drive to analyze the variables in a system to derive the underlying rules that govern its behavior. Systemizing also refers to the drive to construct systems. Systemizing allows one to *predict* the behavior of a system and to control it. I review the evidence indicating that, on average, males spontaneously systemize to a greater degree than do females.[2]

As systemizing is a new concept, it needs a little more definition. By a "system" I mean something that takes inputs and delivers outputs. To systemize, one uses "if–then" (correlation) rules. The brain focuses on a detail or parameter of the system and observes how this varies—that is, it treats a feature of a particular object or event as a variable. Alternatively, a person actively or systematically manipulates a given variable. One notes the effect(s) of performing an operation on one single input in terms of its effects elsewhere in the system (the output). The key data structure used in systemizing is [input–operation–output]. If I do *x*, *a* changes to *b*. If *z* occurs, *p* changes to *q*. Systemizing therefore requires an exact eye for detail.

As shown in table 1-1, the human brain can analyze or construct at least six kinds of systems. Systemizing is an inductive process. One watches what happens each time, gathering data about an event from repeated sampling, often quantifying differences in some variables within the event and observing their correlation with variation in outcome. After confirming a reliable pattern of association—that is, generating predictable results—one forms a rule about how a particular aspect of the system works. When an exception occurs, the rule is refined or revised. Otherwise, the rule is retained. Systemizing works for phenomena that are ultimately lawful, finite, and deterministic. The explanation is exact, and its truth-value is testable. ("The light went on because the switch was in the up position.")

Systemizing is of almost no use for predicting moment-to-moment changes in a person's behavior, but it is our most powerful way of understanding and predicting the law-governed, inanimate universe.

TABLE 1-1

MAIN TYPES OF SYSTEMS

Systems	Examples
Technical	A computer, a musical instrument, a hammer
Natural	A tide, a weather front, a plant
Abstract	Mathematics, a computer program, syntax
Social	A political election, a legal system, a business
Organizable	A taxonomy, a collection, a library
Motoric	A sports technique, a performance, a musical technique

The relevant domains to explore for evidence of systemizing include any fields that are, in principle, rule-governed. Thus, chess and football are good examples of systems. As noted above, systemizing involves monitoring three elements: input, operation, and output. The operation is what was done or what happened to the input in order to produce the output.

So, what is the evidence for a stronger drive to systemize in males?

- *Toy preferences.* Boys are more interested than girls in toy vehicles, weapons, building blocks, and mechanical toys, all of which are open to being "systemized."[3]

- *Adult occupational choices.* Some occupations are almost entirely male. These include metalworking, weapon-making, the manufacture of musical instruments, and the construction industries, such as boatbuilding. The focus of these occupations is on creating systems.[4]

- *Predominantly male disciplines.* Math, physics, computer-science, and engineering all require high systemizing and are largely male-dominated.

- *Test scores.* The SAT Reasoning Test (formerly the Scholastic Aptitude Test and Scholastic Assessment Test), which is administered nationally to college applicants in the United States, is, in part, a test of math skills. Males on average score 50 points higher than females on this portion of the test.[5] Among individuals who score above 700 (out of a possible 800) points, the sex ratio is thirteen to one (men to women).[6]

- *Constructional abilities.* On average, men score higher than women in an assembly task in which people are asked to put together a three-dimensional (3-D) mechanical apparatus. Boys are also better at constructing block buildings from two-dimensional blueprints, and they show more interest than girls in playing with LEGO bricks, which can be combined and recombined into an infinite number of systems. Boys as young as three years of age are also faster at copying 3-D models of outsized LEGO pieces. Older boys, from the age of nine years, are better than girls at imagining what a 3-D object will look like if it is laid out flat, and at constructing a 3-D structure from just an aerial and frontal view in a picture.[7]

- *The water-level task.* Originally devised by the Swiss child psychologist Jean Piaget, the water-level task involves a bottle that is tipped at an angle. Individuals are asked to predict the water level. Women more often draw the water level aligned with the tilt of the bottle and not horizontally, as is correct.[8]

- *The rod-and-frame test.* The rod-and-frame test features a movable rod inside a movable frame. As the frame is moved, the subject is asked to adjust the rod to keep it in a vertical position. A person whose judgment of vertical orientation is influenced by the tilt of the frame is said to be "field-dependent"—that is, his or her judgment is easily swayed by extraneous input in the surrounding context. One who is not influenced by the tilt of the frame is said to be "field-independent." Most studies indicate that females are more field-dependent—that is, women are relatively more distracted by contextual cues, and they tend not to consider each variable within a system separately. They are more likely than men to state erroneously that a rod is upright if it is aligned with its frame.[9]

- *The embedded-figures test.* Attention to relevant detail, which is a general feature of systemizing and clearly a necessary part of it, is superior in males. One measure of this is the embedded-figures test. On average, males are quicker and more accurate than women in locating a target object in a larger, complex pattern.[10] Males, on average, are also better at detecting a particular feature (static or moving).[11]

- *The mental rotation test.* The mental rotation test involves systemizing because it is necessary to treat each feature in a display as a variable that can be transformed (for instance, rotated) and then predict the output, or how it will appear after transformation. Again, men are quicker and more accurate than women in performing the task.[12]

- *Reading maps.* Map-reading is an everyday test of systemizing in that it requires features from 3-D input to be transformed to a two-dimensional representation. In general, boys perform at a higher level than girls in map-reading. Men can also learn a route by looking at a map in fewer trials than women, and they are more successful at correctly recalling details about direction and distance. This observation suggests that men treat features in a map as variables that can be transformed into three dimensions. When children are asked to make a map of an area that they have only visited once, boys' maps have a more accurate layout of the features in the environment. More of the girls' maps make serious errors in the location of important landmarks. Boys tend to emphasize routes or roads, whereas girls tend to emphasize specific landmarks (the corner shop, the park, and so on). These strategies of using directional cues versus using landmark cues have been widely studied. The directional strategy represents an approach to understanding space as a geometric system. Similarly, the focus on roads or routes is an example of considering space in terms of another system, in this case a transportation system.[13]

- *Motoric systems performance.* When asked to throw or catch moving objects (target-directed tasks), such as playing darts or intercepting balls flung from a launcher, males tend to perform better than females. In addition, men are, on average, more accurate than women in their ability to judge which of two moving objects is traveling faster.[14]

- *The Systemizing Quotient.* A questionnaire that has been tested among adults in the general population, the Systemizing Quotient includes forty items that ask about a subject's level of interest in a range of different systems existing in the environment, including technical, abstract, and natural systems. Males score higher than females on this measure.[15]

- *Mechanics test.* The Physical Prediction Questionnaire (PPQ) is based on an established method for selecting applicants to study engineering. The task involves predicting in which direction levers will move when an internal mechanism of cogwheels and pulleys is engaged. Men score significantly higher on this test than women.

Female Advantage in Empathy

We have summarized the evidence for a stronger interest in systems in males, but there is also a body of evidence suggesting that females have a stronger interest in and aptitude for empathy. As summarized below, sex differences of a small but statistically significant magnitude have been found by studies in a number of areas:

- *Sharing and turn-taking.* On average, girls show more concern for fairness in sharing, while boys share less. In one study, boys showed a propensity for competition fifty times greater than that of girls, while girls were twenty times more likely than boys to take turns.[16]

- *Rough-and-tumble play, or "rough-housing."* Boys engage in more wrestling, mock fighting, and other such activities than girls. While often playful, rough-housing can cause injuries or be intrusive, suggesting that higher levels of empathy may tend to discourage it.[17]

- *Responding empathically to the distress of other people.* Girls from the age of one year show greater concern for others through sad looks, sympathetic vocalizations, and comforting behavior than do boys. More women than men report frequently sharing the emotional distress of their friends and demonstrate more comforting behavior, even toward strangers, than men do.[18]

- *Using a "theory of mind."* As early as three years of age, little girls are ahead of boys in their ability to infer what people might be thinking or intending.[19]

- *Sensitivity to facial expressions.* Women are better at decoding nonverbal communication, picking up subtle nuances from tone of voice or facial expression, or judging a person's character.[20]

- *Tests of empathy.* Women score higher than men on questionnaires designed to measure empathic response.[21]

- *Values in relationships.* More women than men value the development of altruistic, reciprocal relationships, which by definition require empathizing. In contrast, more men than women value power, politics, and competition.[22] Girls are more likely to endorse cooperative items on a questionnaire and to rate the establishment of intimacy as more important than the establishment of dominance. In contrast, boys are more likely than girls to endorse competitive items and to rate social status as more important than intimacy.[23]

- *Disorders of empathy.* Disorders such as psychopathic personality disorder and conduct disorder are far more common among males.[24]

- *Aggression.* Aggression can occur only with reduced empathizing. Here again, there is a clear sex difference. Males tend to display far more "direct" aggression (such as pushing, hitting, and punching), while females tend to show more "indirect" (relational, covert) aggression (such as engaging in gossip, excluding others, and making cutting remarks). Engaging in direct aggression may involve a lower level of empathy than engaging in indirect aggression, while indirect aggression may call for better mind-reading skills because its impact is strategic.[25]

- *Murder.* The deliberate taking of life is the ultimate demonstration of a lack of empathy. Daly and Wilson analyzed homicide records dating back over seven hundred years, from a range of different societies. They found that "male-on-male" homicide was thirty to forty times more frequent than "female-on-female" homicide.[26]

- *Establishment of "dominance hierarchies."* Males are quicker than females to establish forms of social organization in which members compete over resources by means of aggression. Typically, a dominance hierarchy is established by one or more individuals pushing others around to become the leaders, which in part may reflect lower empathizing skills.[27]

- *Language style.* Girls' speech is more cooperative, reciprocal, and collaborative than that of boys. In concrete terms, this difference is reflected in girls' ability to continue a conversational exchange with a partner for a longer period. When girls disagree, they are more likely to express their differing opinions sensitively, in the form of questions rather than assertions. Boys' talk is more "single-voiced discourse"—that is, the speaker presents only his own perspective. The female speech style is more "double-voiced discourse"—a girl will spend more time negotiating with her partner, trying to take the other person's wishes into account.[28]

- *Language abilities.* Females have been shown to have better language skills than males. It seems likely that good empathizing would promote language development,[29] and vice versa, so these factors may not be independent.

- *Talk about emotions.* Women's conversations involve much more talk about feelings than men, while men's conversations tend to be more object- or activity-focused.[30]

- *Parenting style.* Fathers are less likely than mothers to hold their infants in a face-to-face position. Mothers tend to go along with their children's choices in play, while fathers are more likely to impose their own choices. Also, mothers more often fine-tune their speech to match their children's understanding.[31]

• *Face preference and eye contact.* From birth, females look longer at faces, particularly at people's eyes, whereas males are more likely to look at inanimate objects.[32]

Culture and Biology

At one year of age, boys strongly prefer to watch a video of cars going past, an example of predictable mechanical systems, than to watch a film showing a human face. Little girls show the opposite preference. Girls also engage in more eye contact than boys at this age.[33] Some investigators argue that differential socialization may cause such sex differences, even at a very early age.

While evidence does exist for socialization contributing to these differences, this is unlikely to be a sufficient explanation. Connellan and colleagues have shown that among *one-day-old* babies, boys look longer at a mechanical mobile, which is a system with predictable laws of motion, than at a person's face, an object that is next to impossible to systemize. One-day-old girls show the opposite profile.[34] These sex differences are, therefore, present earlier in life than can be plausibly explained by socialization, raising the possibility that, while culture and socialization may to some extent determine the development of a brain prone to a stronger interest in systems or empathy, biology may also play a part. Evidence supporting both cultural determinism and biological determinism is ample.[35] For example, one's score on the Systemizing Quotient (SQ) questionnaire is positively correlated with the prenatal level of testosterone.[36]

Conclusions

We have reviewed much evidence suggesting significant sex differences in the drive to systemize and empathize. While on some tests this is expressed in terms of ability, my own view is that these differences are fundamentally a reflection of *drives* or *interests* rather than ability per se. That is, on average, more boys than girls are attracted to systems from an early age, and this difference leads more boys to pursue activities (such as math or music or skateboarding) that involve systemizing. Increased practice can lead to stronger ability, but it remains plausible that these are primarily differences in personality, with differences in ability being secondary. Equally, we have reviewed evidence that, on average, more girls than boys are attracted to people and the emotional lives of others, which involves empathizing.

The causes of these fundamental differences remain unclear, but over and above the role of experience and the postnatal environment (including differences in socialization), candidates for prenatal biological factors that may be implicated include both genetic differences and testosterone levels in utero.[37] We can find another clue that systemizing and empathizing may have a partly genetic basis in the fact that in the neurodevelopmental condition of autism, which is genetic, the drive to systemize is even stronger than in the general population, while empathy is impaired. Indeed, it is possible that autism exemplifies "extreme maleness."[38]

The research reviewed above suggests we should not expect the sex ratio in occupations such as math or physics ever to be fifty-fifty if the workplace is left simply to reflect the numbers of applicants of each sex who are drawn to such fields. If we want a particular field to have an equal representation of men and women, which I think may be desirable for reasons other than scientific, we need to put in place social policies that will produce that outcome.

Finally, and most importantly, the research teaches us that there is no scientific justification for stereotyping, since none of the studies allows one to predict an individual's aptitudes or interests on the basis of his or her sex. This is because—at risk of repetition—the studies only capture differences between groups on average. Individuals are just that: They may be typical or atypical for their group. Prejudging an individual on the basis of his or her sex is, as the word "prejudge" suggests, mere prejudice.

Notes

1. For this second argument, I am grateful to Steve Pinker both for pointing it out and for giving permission to reproduce figures 1-1, 1-2, and 1-3 from his files.

2. Baron-Cohen et al. 2002.

3. Jennings 1977.

4. Geary 1998.

5. Benbow and Stanley 1983.

6. Geary 1996.

7. Kimura 1999.

8. Wittig and Allen 1984.

9. Witkin et al. 1954.

10. Elliot 1961.

11. Voyer et al. 1995.

12. Collins and Kimura 1997.

13. Galea and Kimura 1993.

14. Schiff and Oldak 1990.

15. Baron-Cohen et al. 2003.

16. Charlesworth and Dzur 1987.

17. Maccoby 1999.

18. Hoffman 1977.

19. Happe 1995.

20. Davis 1994.

21. Baron-Cohen and Wheelwright 2004.

22. Ahlgren and Johnson 1979.

23. Knight et al. 1989.

24. Dodge 1980; Blair 1995.

25. Crick and Grotpeter 1995.

26. Daly and Wilson 1988.

27. Strayer 1980.

28. Smith 1985.

29. Baron-Cohen et al. 1997.

30. Tannen 1990.

31. Power 1985.

32. Connellan et al. 2001.

33. Lutchmaya and Baron-Cohen 2002.

34. Connellan et al. 2001.

35. Eagly 1987; Gouchie and Kimura 1991.

36. Lutchmaya et al. 2002. For a review of the evidence for the biological basis of sex differences in the mind, see Baron-Cohen 2003.

37. Bailey, Bolton, and Rutter 1998.

38. Baron-Cohen 2003.

References

Ahlgren, A., and D. W. Johnson. 1979. Sex Differences in Cooperative and Competitive Attitudes from the 2nd to the 12th Grades. *Developmental Psychology* 15: 45–49.

Bailey, A., P. Bolton, and M. Rutter. 1998. A Full Genome Screen for Autism with Evidence for Linkage to a Region on Chromosome 7q. *Human Molecular Genetics* 7 (3). 571–78.

Baron-Cohen, S. 2003. *The Essential Difference: Men, Women and the Extreme Male Brain.* London: Penguin.

———, D. Baldwin, and M. Crowson. 1997. Do Children with Autism Use the Speaker's Direction of Gaze (Sdg) Strategy to Crack the Code of Language? *Child Development* 68 (1): 48–57.

———, Jennifer Richler, Dheraj Bisarya, Nhishanth Gurunathan, and Sally Wheelwright. 2003. The Systemising Quotient (Sq): An Investigation of Adults with Asperger Syndrome or High Functioning Autism and Normal Sex Differences. *Philosophical Transactions of the Royal Society* 358 (1430): 361–74.

———, and S. Wheelwright. 2004. The Empathy Quotient (Eq). An Investigation of Adults with Asperger Syndrome or High Functioning Autism, and Normal Sex Differences. *Journal of Autism and Developmental Disorders* 34 (2): 163–75.

———, Sally Wheelwright, John Lawson, Rick Griffin, and Jacqueline Hill. 2002. The Exact Mind: Empathising and Systemising in Autism Spectrum Conditions. In *Blackwell Handbook of Childhood Cognitive Development*, ed. U. Goswami. Oxford: Wiley-Blackwell.

Benbow, C. P., and J. C. Stanley. 1983. Sex Differences in Mathematical Reasoning Ability: More Facts. *Science* 222: 1029–31.

Blair, R. J. 1995. A Cognitive Developmental Approach to Morality: Investigating the Psychopath. *Cognition* 57 (1): 1–29.

Charlesworth, W. R., and C. Dzur. 1987. Gender Comparisons of Preschoolers' Behavior and Resource Utilization in Group Problem-Solving. *Child Development* 58 (1): 191–200.

Collins, D. W., and D. Kimura. 1997. A Large Sex Difference on a Two-Dimensional Mental Rotation Task. *Behavioral Neuroscience* 111 (4): 145–49.

Connellan, J., S. Baron-Cohen, S. Wheelwright, A. Ba'tki, and J. Ahluwalia. 2001. Sex Differences in Human Neonatal Social Perception. *Infant Behavior and Development* 23: 113–18.

Crick, N. R., and J. K. Grotpeter. 1995. Relational Aggression, Gender and Social-Psychological Adjustment. *Child Development* 66 (3): 710–22.

Daly, M., and M. Wilson. 1988. *Homicide.* New York: Aldine de Gruyter.

Davis, M. H. 1994. *Empathy: A Social Psychological Approach, Social Psychology Series.* Boulder, CO: Westview Press.

Dodge, K. 1980. Social Cognition and Children's Aggressive Behaviour. *Child Development* 51 (1): 162–70.

Eagly, A. H. 1987. *Sex Differences in Social Behavior: A Social-Role Interpretation.* Hillsdale, N.J.: Erlbaum.

Elliot, R. 1961. Interrelationship among Measures of Field Dependence, Ability, and Personality Traits. *Journal of Abnormal and Social Psychology* 63: 27–36.

Galea, L. A. M., and D. Kimura. 1993. Sex Differences in Route Learning. *Personality and Individual Differences* 14 (1): 53–65.

Geary, D. 1996. Sexual Selection and Sex Differences in Mathematical Abilities. *Behavioral and Brain Sciences* 19 (2): 229–84.

———. 1998. *Male, Female: The Evolution of Human Sex Differences*. Washington, D.C.: American Psychological Association.

Gouchie, C., and D. Kimura. 1991. The Relationship between Testosterone Levels and Cognitive Ability Patterns. *Psychoneuroendocrinology* 16 (4): 323–34.

Happe, F. 1995. The Role of Age and Verbal Ability in the Theory of Mind Task Performance of Subjects with Autism. *Child Development* 66 (3): 843–55.

Hoffman, M. L. 1977. Sex Differences in Empathy and Related Behaviors. *Psychological Bulletin* 84: 712–22.

Jennings, K. D. 1977. People versus Object Orientation in Preschool Children: Do Sex Differences Really Occur? *Journal of Genetic Psychology* 131 (1): 65–74.

Kimura, D. 1999. *Sex and Cognition*. Cambridge, Mass.: MIT Press.

Knight, G. P., R. A. Fabes, and D. A. Higgins. 1989. Gender Differences in the Cooperative, Competitive, and Individualistic Social Values of Children. *Motivation and Emotion* 13: 125–41.

Lutchmaya, S., and S. Baron-Cohen. 2002. Human Sex Differences in Social and Non-Social Looking Preferences at 12 Months of Age. *Infant Behavior and Development* 25 (3): 319–25.

Lutchmaya, S., S. Baron-Cohen, and P. Raggatt. 2002. Foetal Testosterone and Eye Contact in 12 Month Old Infants. *Infant Behavior and Development* 25: 327–35.

Maccoby, E. 1999. *The Two Sexes: Growing up Apart, Coming Together.* Cambridge, Mass.: Belknap Press of Harvard University Press.

Power, T. G. 1985. Mother– and Father–Infant Play: A Developmental Analysis. *Child Development* 56 (6): 1514–24.

Schiff, W., and R. Oldak. 1990. Accuracy of Judging Time to Arrival: Effects of Modality, Trajectory and Gender. *Journal of Experimental Psychology, Human Perception and Performance* 16 (2): 303–16.

Smith, P. M. 1985. *Language, the Sexes and Society.* Oxford: Basil Blackwell.

Strayer, F. F. 1980. Child Ethology and the Study of Preschool Social Relations. In *Friendship and Social Relations in Children*, ed. H. C. Foot, A. J. Chapman, and J. R. Smith. New York: John Wiley.

Tannen, D. 1990. *You Just Don't Understand: Women and Men in Conversation.* New York: William Morrow.

Voyer, D., S. Voyer, and M. Bryden. 1995. Magnitude of Sex Differences in Spatial Abilities: A Meta-Analysis and Consideration of Critical Variables. *Psychological Bulletin* 117 (2): 250–70.

Witkin, H. A., H. B. Lewis, M. Hertzman, K. Machover, P. Bretnall Meissner, and S. Wapner. 1954. *Personality through Perception*. New York: Harper and Brothers.

Wittig, M. A., and M. J. Allen. 1984. Measurement of Adult Performance on Piaget's Water Horizontality Task. *Intelligence* 8: 305–13.

2

Gender, Math, and Science

Elizabeth S. Spelke and Katherine Ellison

In all known cultures, and at all times in human development, gender has mattered. Three-month-old infants look differently at male and female faces, preferring faces of the gender of their primary caregivers.[1] Young children tend to choose other children of the same gender as friends[2] and favor objects and activities that children of their own gender endorse.[3] Young adults tend to gravitate to work practiced by people of their own gender,[4] and older adults tend to evaluate more positively job applicants of the gender that predominates in their fields.[5] It is not surprising, therefore, that professions have tended to be segregated by gender all over the world, and at all times in history.

Some of this clustering must depend on social and historical factors, because it varies over time and across cultures. Nursing and accounting were once male-dominated professions in the United States, for example, but now most American nurses and accountants are female. Some of the clustering might depend, however, on biological factors. In most societies, academic occupations tend to attract disproportionate numbers of males, especially in the sciences. It is reasonable to ask whether men and women differ in the cognitive aptitudes and motivations that make for success in these fields.

These are questions that motivated the much-discussed remarks of Harvard University's former president, Lawrence Summers, in 2005, suggesting that males have a genetic advantage in science and mathematics that could account for their high representation in those fields.[6] Following Summers, we ask in this chapter whether this is true. In particular, we consider three hypotheses that he advanced.[7] Summers conjectured, first, that males have a higher intrinsic aptitude for math and science, both on average and at the high end of ability. Second, he suggested that males are more interested in objects and mechanics and are predisposed to work harder and more intensely at math and science, among other pursuits. Third, he argued that gender discrimination is countered by market forces and therefore plays little role in accounting for the dearth of women in high-level professions. Our review of research provides evidence against all three of these hypotheses. The primary causes of the gender gap in academic science and mathematics, we suggest, are social and historical rather than genetic and psychological.

Gender Differences in Intrinsic Aptitude for Math and Science?

Symbolic mathematics and science are recent achievements on an evolutionary time scale. Recursive natural number systems emerged only several thousand years ago, and they may not be universal across today's human cultures.[8] Geometric mapmaking is even more recent, and the formal unification of numbers and geometry is less than four hundred years old.[9] Any capacity that is specifically adapted for formal mathematics and science, therefore, cannot have evolved in humans; there are no genes for calculus or thermodynamics. When children learn science and mathematics, and when adults practice in these fields, they bring to bear systems of the mind and brain that evolved to serve other functions.

Research in cognitive psychology and neuroscience provides converging evidence for three core systems at the foundations of mathematics and science: a system for representing small, exact numbers of objects (up to three); a system for assessing and comparing large, approximate numerical magnitudes (for example, about twenty); and a system for detecting geometric properties and relationships (especially Euclidean distance and angle).[10] Each of these core systems emerges in human infancy and continues to function in children and adults in widely differing cultures.[11] Each, moreover, is shared by nonhuman animals, including primates, other mammals, and even birds, and therefore has deep roots in cognitive evolution.[12] In behavioral experiments, these systems have been found to guide adults' intuitive reasoning about object mechanics,[13] mental arithmetic,[14] and geometrical relationships.[15] In experiments using functional brain imaging, they show activation when adults or children solve problems in symbolic mathematics.[16]

Furthermore, neurological patients with impairment of these systems show associated impairments in symbolic mathematics and spatial cognition.[17]

Most importantly, each of these systems supports young children's learning of formal mathematics and science. Children use their core representations of objects to learn the detailed mechanical properties of objects and the system of verbal counting.[18] They use their representations of large, approximate numbers to solve problems in symbolic arithmetic[19] and master its logical properties, such as the inverse relationship between addition and subtraction.[20] And they use the system of geometric representation to make sense of symbolic maps and higher mathematics.[21]

Although the three systems at the core of mathematical and scientific thinking are relatively independent of one another at the start of life, they become linked together during childhood. The linkages that support some of the most important academic skills emerge before children begin school. Children master verbal counting, at about age four, by connecting their representations of small, exact numbers and large, approximate numbers so as to construct the system of natural numbers.[22] They also begin to link their representations of objects and geometry: Four-year-olds detect the geometric relationships among objects in a set and use those relationships to understand maps.[23] These linkages provide highly useful and versatile tools for mastering school mathematics and science.[24]

Armed with these findings, psychologists can investigate Summers's first hypothesis. We can ask whether boys outperform girls at tasks that tap any of the three core systems—that is, do boys show greater aptitude, either on average or at high levels, for representing small numbers of objects, comparing and operating on approximate numerical magnitudes, or detecting the geometric properties of surface layouts and visual forms? Moreover, we can ask whether boys have superior abilities to link the systems together so as to count and navigate by maps. Let us consider the evidence bearing on these questions.

Object Representations. Since the pioneering research of Swiss psychologist Jean Piaget, begun more than fifty years ago, psychologists have investigated the capacities of human infants to represent objects. Experiments using simple behavioral methods or methods of functional brain imaging reveal that infants perceive objects that are visible, remember objects that are hidden, and make basic inferences about objects' behavior. Infants infer, for example, that a moving block will not pass through a solid barrier, that it will move when hit but will not move spontaneously, and that it will continue to exist when it moves out of view.[25] Importantly, infants' object representations have critical limits; for example, infants can represent up to three or four objects at a time, but not more.

One experiment, by Feigenson and others, serves as an example. In this study, ten-month-old infants were presented with two boxes, placed apart from one another and beyond their reach. While each infant watched, the experimenter placed two crackers in one box and three crackers in the other, one at a time. Then the experimenter looked down and encouraged the infant to crawl toward a box and "get the crackers." Infants tended reliably to crawl toward the box with three rather than two crackers. Subsequent experiments showed that their choice depended on representations of the hidden objects in each box, rather than representations of visible properties of the events (such as a hand entering the box). Moreover, experiments showed that infants chose the larger number only when both numbers were small: up to three.[26]

These findings allow us to ask whether male infants are better at representing objects than female infants. In an important review covering the first thirty years of research on this topic, Maccoby and Jacklin concluded that male and female infants and toddlers show equal interest in objects and equal abilities to represent them.[27] More recent studies of object representation in infants also tend to show no sex differences, although one, comparing the number of objects that infants can keep track of using Feigenson's method, found a small female advantage.[28] Studies of infants, therefore, provide no evidence that males have greater intrinsic aptitude for reasoning about the physical world.

This negative conclusion can be questioned on three grounds.[29] First, studies of infants tend to reveal what infants do spontaneously, but not what they are capable of doing when pressed. It is possible that one gender would show greater interest in objects and better understanding of their behavior under conditions of high motivation and task demands. Second, such studies tend to focus on the capacities infants show on average, but they rarely test enough of them to focus on performance at high levels. Third, these studies can reveal genetically determined capacities that emerge early in development, but not those that emerge later. To address these points, we need methods to study the core system of object representation at later ages, under conditions of greater task demands, and in larger populations.

One useful task for this purpose is multiple-object tracking.[30] Participants are shown an array of indistinguishable discs on a computer screen and asked to attend to a subset of the discs that are indicated to them. Then all the discs begin to move independently, while the participants continue to focus their attention on the previously designated subset. After some seconds, the discs stop moving, and participants report the positions of those in the tracked subset. Experiments using this method provide evidence that adults engaged in multiple-object tracking show the same abilities, and limits, as infants engaged in tasks of object representation; for example, adults cannot track more than three or four objects reliably.[31] Moreover, the task requires high levels of effort. If males have a genetic advantage at representing objects, therefore, they should track objects more accurately. Because tracking performance has been shown to be highly malleable by experience with video games in adolescence and adulthood,[32] we were especially interested in the performance of children on this task, spanning the ages when children first begin their formal education.

To test for sex differences in young children, we created a child-friendly version of the multiple-object tracking task (shown in figure 2-1a) and administered it to children ages four to eight years. Participants saw an array on a computer screen of eight black dots described as ladybug eggs, and they were instructed to keep track of a subset of two to six eggs while all the eggs moved independently around the screen. After ten seconds of motion, the eggs stopped moving, two eggs turned into ladybugs, and each child indicated which of the two ladybugs was in his or her tracked set. Children of all ages appeared highly engrossed by this task, and they performed quite well. When we compared performance of boys and girls, both in the full sample and at the highest level, we found no sex differences favoring males (see figure 2-1b). Boys and girls were equally able to perceive, attend to, and remember objects in this challenging task.

Numerical Representations. Research over the past thirty years provides evidence that mathematical thinking depends, in part, on a universal, genetically based capacity to represent and reason about numerical magnitudes.[33] This capacity gives rise to numerical representations with four fundamental properties:

FIGURE 2-1
DISPLAYS AND FINDINGS FOR THE TEST OF
MULTIPLE-OBJECT TRACKING

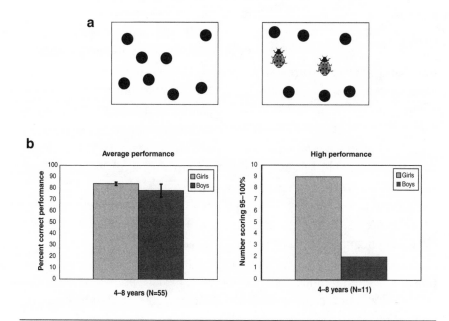

SOURCE: Authors' diagram.

- They are approximate and subject to a ratio limit on "discriminability" (that is, for example, twenty can be discriminated from forty about as easily as ten from twenty).

- They are abstract and independent of modality and format (twenty simultaneously presented visible dots can be compared to thirty dots about as easily as they can be compared to thirty sequentially presented sounds).

- They are ordered (thirty is represented as *more* than twenty).

- They can be transformed in accordance with the laws of arithmetic (ten dots can be added to ten dots to yield a representation of approximately twenty dots).

The sense of number is present in human infants, as well as in preschool children and adults.[34] A simple experiment conducted by Barth and others on preschool children illustrates its four key properties.[35] In this study, children with no formal instruction in arithmetic were shown animated computer displays involving visual arrays of dots, auditory sequences of tones, or both. In some cases, one array or sequence was followed by another, and children were asked to compare them. In others, three arrays or sequences were presented, and children were asked to add the first two sets together and compare the sum to the third set. Figure 2-2a shows an example of a visual addition trial. On such visual comparison trials, the middle event was eliminated, and the two sets of black dots appeared as a single quantity. On cross-modal trials, a black-dot array was replaced by a sequence of sounds. On all of these tasks, preschool children showed the same patterns of performance as adults tested with the same types of displays. Moreover, children's performance paralleled adults' performance on similar tasks in which numbers appeared as Arabic digits, tapping understanding of symbolic arithmetic.[36]

Given these findings, we can ask whether male and female infants or preschool children differ from each other in their respective nonsymbolic numerical abilities. In the large body of experiments on number sense in human infants, no consistent sex differences have been reported. Because studies of infants' number sense are subject to the same criticisms raised for studies of infants' representations of objects, however, it is worth examining in more detail the task used by Barth and colleagues and shown in figure 2-2a.[37] This task, which is highly engaging to preschool children, has revealed differences among them that are consistent and predictive of academic success. Furthermore, it has been presented to a sample of children large enough to test not only for sex differences in average performance, but in high performance. As figure 2-2b indicates, these tests yield no sex differences at either level. Boys and girls appear to have equal intrinsic aptitude for numerical representations and reasoning.

Geometrical Reasoning. Studies of geometrical representations and reasoning have a rich history in experimental psychology and neuroscience, and their findings have figured prominently in discussions of cognitive sex differences.[38] Studies of infants and young children reveal an early-developing

FIGURE 2-2

DISPLAYS AND FINDINGS FOR THE TEST OF
NONSYMBOLIC ADDITION AND COMPARISON

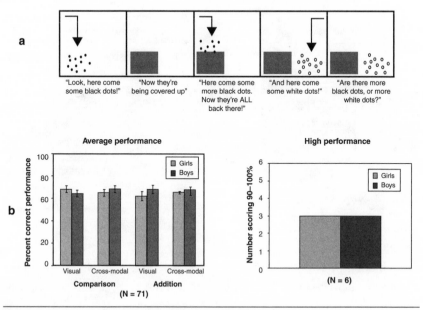

SOURCE: Authors' diagram.

sensitivity to the geometry of both surface layout and visual patterns—a sensitivity that is shared by other animals, develops in animals under strong genetic constraint, and is preserved in adults in diverse cultures.[39]

Two tasks have been especially useful in revealing this sensitivity. Using the first, sensitivity to the geometry of the surrounding spatial layout has been tested in animals by showing a hungry animal the location of hidden food in a distinctively shaped chamber, disorienting the animal, and then allowing the animal to search for the food. In this task, animals will only find the hidden food if they can reorient themselves. A wide variety of animals reorient themselves and carry out the task by encoding, detecting, and remembering the shape of the surrounding surface layout.[40]

Studies using a variation of this task reveal the same ability in human adults and infants.[41] In figure 2-3, children are introduced into a rectangular room with no landmark objects or surface markings. They watch as a

FIGURE 2-3

FINDINGS FOR THE TEST OF SENSITIVITY TO
SURFACE LAYOUT GEOMETRY

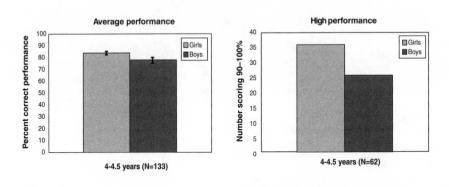

SOURCE: Authors' diagram.

toy is hidden in a corner, and then turn slowly with eyes closed to induce a state of disorientation but not dizziness; then they search for the toy. Sensitivity to geometry is inferred from the location at which the child searches for the toy. (The possibility of choosing one of the two geometrically appropriate corners by chance is 50 percent.) In this task, both adults and children show high sensitivity to geometry, directing their search to the two geometrically appropriate corners.

The second task has been used by Dehaene and others to test sensitivity to geometry in visual forms in children and adults living in two cultures in the United States and Brazil: in the metropolitan Boston area and in remote villages in the Amazon, respectively.[42] For this test, participants in each trial view six geometric figures that differ in size and orientation. Five of the figures share a single property not shared by the sixth, and participants are instructed to find the deviant figure (see figure 2-4a; the possibility of choosing the correct answers by chance is 17 percent). After many trials testing for a variety of geometric properties with the Boston and Amazonian groups, two striking findings emerged. First, the overall level of performance was influenced by culture and/or education: Boston adults performed better than, collectively, Boston children, Amazonian adults

FIGURE 2-4
DISPLAYS AND FINDINGS FOR THE TEST OF SENSITIVITY
TO GEOMETRY IN VISUAL FORMS

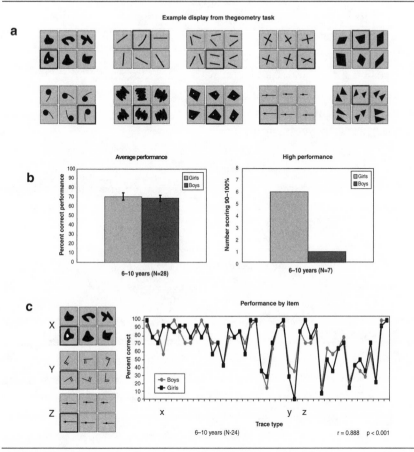

SOURCE: Authors' diagram.

(who lacked any formal education), and Amazonian children; performance of the last three groups did not differ. Second, all groups performed well above chance but far from perfectly, and they showed a highly similar performance profile: Items that were relatively more difficult for Boston adults were also more difficult for participants in each of the other groups. These findings provide evidence that a core system for representing geometric properties and relationships develops universally in humans, independent of formal instruction or other culture-specific experiences.

The two tasks just described can be used to test whether males show greater core spatial abilities than females. Because the reorientation task has been used with infants in numerous experiments, investigators have asked whether male infants reorient by geometry more effectively than female infants. No study has found such a sex difference.[43] To address Pinker's suggestions as described earlier, we have also compared the performance of boys and girls at older ages and with larger samples, focusing both on average performance and on the subset of children showing the highest performance. As figure 2-3b shows, we found no significant sex differences favoring males on this navigation task.

The geometrical form task of Dehaene and others serves as a further test of possible differences in spatial ability between boys and girls.[44] To this end, we administered the task to groups of children and adults in Boston and compared the performances of the two sexes, both on average and at the highest levels. The findings of this analysis were clear and convergent with those of the previous test: no sex differences favoring males on average or at the highest levels of performance (see figure 2-4b).[45]

Although boys do not outperform girls on this task, it remains possible that a narrower pattern of sex differences will appear. Maybe boys are especially good at detecting certain kinds of geometrical properties, and girls are better at detecting others. To test this hypothesis, we compared the detailed performance of the two sexes across the forty-five items in this test. The graph in figure 2-4c shows the proportion of boys and girls performing correctly on each item, with the items labeled x–z indicating marginally significant sex differences favoring girls (x and z) or boys (y). The comparison revealed not only similar overall performance by boys and girls on this task, but highly convergent performance profiles. No item on the test was easy for one sex but difficult for the other. With proper statistical corrections, no items yielded significant sex differences, and only three of the forty-five items showed hints of a difference. Two of the three favored girls.

The third, testing the geometric property of chirality—that is, the property that distinguishes an object from its mirror image—favored boys. To detect chiral objects, people often rotate one object into the same orientation as the other. Males tend to outperform females on mental rotation tasks,[46] even in infancy,[47] and this sex difference often is taken as evidence that males have higher overall spatial abilities. Our task replicated this finding, but our results overall suggest a different conclusion. Males' advantage in detecting chirality is not indicative of a general male advantage in geometrical reasoning. To the contrary, the core geometrical abilities of males and females are highly similar.

The Construction of Natural Numbers. Because the above findings suggest that boys and girls are equally endowed with three core capacities at the foundations of mathematics, we now ask whether preschool children differ in their abilities to link core representations and develop two symbolic cognitive skills that may be crucial for the mastery of academic science and mathematics: verbal counting and symbolic map-reading.

Between the ages of two and four years, most children come to master the meanings of number words and the logic of the verbal counting routine.[48] Mastery proceeds in a regular order. First children learn the ordered list of counting words as a meaningless sequence. Then they learn the meaning of *one*, then *two*, then *three*. Some continue learning further number words one by one, typically in order. Finally, however, children induce the counting rule: They come to realize that each word in the list refers to a set containing one more object than the set designated by the previous word. Children vary in the speed with which they learn number words and make the critical inductive leap, and this variation is predictive of their successful mastery of the elementary school mathematics curriculum.[49]

The most popular task used to assess children's progressive understanding of counting is the "give a number" task.[50] Children are presented with a large set of objects (for example, twelve toy fish scattered outside a toy pond) and are asked to count them. By age three, most children respond by reciting the number words in correct order up to *ten*. Then children are asked to produce a specific number of objects (for example, the experimenter asks the child to "put three fish in the pond"). At three years, most produce the correct number when asked for "one" or "two" but simply grab a handful of objects (more than two) when asked for higher numbers in their counting list. Based on his or her performance with different requested numbers, each child's point of progression through the process of learning counting can be assessed. As figure 2-5 indicates, boys and girls learn counting at roughly equal rates, showing no male advantage either on average or among the best performers (those who have mastered the largest number words).

FIGURE 2-5

FINDINGS FOR THE TEST OF MASTERY OF
NUMBER WORDS AND VERBAL COUNTING

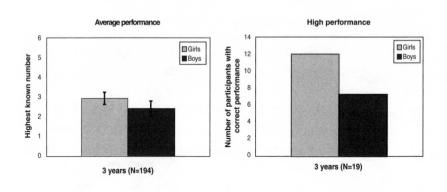

SOURCE: Authors' diagram.

Map-Reading. Children typically begin to understand and use maps at about four years of age, and precursors to this ability may be found at younger ages.[51] Although most studies present children with maps whose colors and shapes resemble those of the layout of objects they depict, a recent experiment reveals that four-year-old children can also navigate by purely geometric maps, with no instruction or feedback given during the task.[52]

In this task, children view a simple map depicting three geometric forms in a triangular or linear arrangement while facing away from an array of three containers forming a similar arrangement, but twelve times larger and at a different orientation relative to the map. On each trial, the experimenter points to a single location on the map and asks the child to place an object in the corresponding container in the array. Across trials, the nature of the arrangement (isosceles triangle, right triangle, or line) and its orientation vary. Sensitivity to geometry is reflected in the proportion of objects placed at the geometrically specified locations. Children performed this task well above chance (which is 33 percent of the objects placed correctly), yet with considerable variability. Once again, girls performed as well as boys on average, and girls were at least as likely as boys to score at the highest levels (see figure 2-6).[53]

FIGURE 2-6

FINDINGS FOR THE TEST OF MAP-BASED NAVIGATION

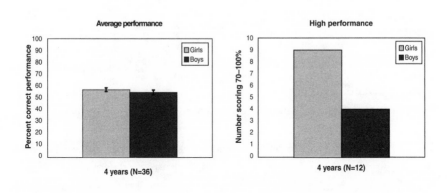

SOURCE: Authors' diagram.

Summary: No Male Advantage in Intrinsic Aptitude. Research on cognitive psychology and cognitive neuroscience provides evidence for three core systems at the foundations of mathematics and science. Core systems for representing objects, numbers, and geometry emerge in human infants, guide children's learning over the preschool years, and support later mastery of formal science and mathematics. In no case have male infants or children been found to have a general advantage over females in any of these core domains. Male and female infants and children perform equally well on tasks tapping these core representations, both on average and at high levels of achievement.

Basic capacities to link these core systems emerge in the preschool years, and they allow children to develop new cognitive skills that are crucial for later academic success. One skill, verbal counting, provides a foundation for learning arithmetic. Another, visual-symbol understanding, provides a foundation for abstract learning in general. Girls and boys also show equal mastery of verbal counting and of one of the earliest visual symbolic skills—use of geometric maps—both on average and at high levels.

None of these findings supports Summers's conjecture of a male advantage in intrinsic cognitive aptitude for mathematics and science.

Gender Differences in Intrinsic Motivation
for Mathematics and Science?

Summers suggested that two genetically determined motivational differences between males and females may predispose the former to succeed in mathematics and science. First, boys may tend to be inherently more interested in the physical world, whereas girls are more interested in the social world.[54] Second, men may be more motivated to work uninterruptedly for long hours, thereby accepting more readily a career that places large demands on their time.

The self-reported interests and preferences of adults are consistent with these conjectures. More females than males appear to be deterred by the long work hours of many successful scientists and engineers, and more gifted female students express an interest in professions involving work with people.[55] The expressed preferences and choices of adults cannot, however, be presumed to reflect genetic differences between the sexes. Work patterns vary widely across cultures and over historical time. In any society and at any age, people tend to aspire to professions that they view as attainable and appropriate; often, these are the professions that are practiced by other people like themselves.[56] As the gender composition of a profession changes, therefore, so does the gender composition of the students who seek to join it. When most U.S. physicians were male two generations ago, for example, so were most applicants to U.S. medical schools. Today, as many more women have come to practice medicine, the numbers of female students expressing an interest in this field have soared. Given the low numbers of senior women physicists and mathematicians today, it is likely that this social force now operates to reduce female students' interest in these fields. Nevertheless, the possibility exists that genetically determined gender differences in motivation contribute to the gender gap in mathematics and science.

To test this possibility, motivational variables must be identified that are genetically determined, in part, and invariant across human cultures—core motivational systems that parallel the core cognitive systems described in the previous section. Moreover, tests for these variables must be administered to children, and performance on them must be related to later academic outcomes. If a given motivational variable predisposes people to successful careers in math or science, then children who score higher on this variable should show greater achievement in academic mathematics and science. Finally, these tests must be administered to young boys and girls, whose performance can be compared.

In the fields of developmental and educational psychology, few tests of temperament or motivation meet these conditions. Although males and females show reliable differences on some motivational and temperamental variables, most motivational patterns are not stable within a person (for example, the same person may be hardworking at some times but not at other times) or social context (for example, the same person may work hard at sports but not at academics), and many differ as a function of culture or experience.[57] Temperamental variables that both vary by sex and remain stable as people develop[58] tend not to be strongly predictive of academic success. For example, males typically are found to be more aggressive than females, and to display different patterns of competition and cooperation.[59] Because the practice of science is spurred by both competition and collaboration, however, it is not clear whether these differences yield differential advantages for one sex. As a second example, it has been reported that women have greater empathy and interest in social relationships than men.[60] There is no evidence, however, that greater empathy leads to lesser success in mathematics or science. Most research on sex differences in motivation, therefore, does not speak to Summers's hypothesis.

Nevertheless, psychologists have investigated one personal characteristic that meets all of the requirements mentioned above: self-regulation. The development of self-regulation has been studied intensively since the pioneering research of Walter Mischel that began many decades ago.[61] Mischel devised a simple test of self-regulation that can be applied to both adults and young children. Participants are given a choice between a small reward now and a larger reward at a later time. For example, a child might be given the option of eating a single marshmallow without delay or receiving a plate of marshmallows at an unspecified later time. The experimenter leaves one marshmallow with the child, exits the room, and measures how long the child waits before consuming it and forgoing a larger, future treat.

Research reveals large individual differences on such tasks, at all ages tested. Moreover, the individual differences are stable over development: Children who show the greatest capacity for self-regulation at young ages will continue to do so at later ages.[62] Most important, these individual differences, measured in childhood, predict later academic success, not only in elementary school but in high school and college. In a recent study, delay of gratification predicted academic success better than a standard measure of IQ.[63] Even after college, people who delayed gratification longer as children tended to be more successful in their careers.[64] Success in academic science and mathematics requires major investments of time and effort in one's studies, and sacrifices of present pleasures for future rewards. If males do, indeed, have a genetic advantage in motivation for science, we might expect them to perform better than females on tests of this motivational pattern.

They do not. Most studies of self-regulation report no significant sex differences in children or adults. A recent meta-analysis did find a small sex difference that was stable over age, but the difference favored females.[65] Girls were slightly more likely than boys to demonstrate the only motivational pattern that has been shown to be developmentally invariant, robust over variations in experience, and predictive of later success in academic pursuits.

Toward an Understanding of the Gender Gap

The research described above supports two conclusions. First, boys and girls are equally endowed with the core cognitive abilities that we harness when we learn and reason about math and science. Second, boys and girls are equally likely to show self-regulatory abilities that predict academic success. In light of these conclusions, it should not be surprising that female students today perform at least as well as male students in all school subjects, including math and science, through high school and college.[66] All these findings undermine the hypothesis that males are intrinsically better suited to math and science.

These findings, however, raise two questions. First, if males and females have equal cognitive and motivational predispositions for academic careers in math and science, why are there more men than women in these fields? This question is particularly acute in light of Summers's third hypothesis regarding sex differences: If no sex differences in abilities exist, then market forces should tend to equalize the sex ratio as employers compete for the most able candidates. Second, if males and females are so convergent in their cognitive capacities and motivational patterns, why do so many psychologists and laypeople believe otherwise? Briefly, we consider two partial answers to these questions, which tend to undermine Summers's third hypothesis. First, humans are predisposed to divide the social world by gender, whether that division is warranted or not. Second, humans are predisposed to naturalize their current social arrangement, overemphasizing its biological roots.

When a new baby is born into a family or community, one question towers over all others: Is it a boy or a girl? When adults encounter new people, our attention to most of their attributes varies by social context, even for genetically determined attributes such as race, but our encoding of gender appears to be mandatory.[67] Adults are not alone in their preoccupation with gender. As we noted, infants as young as three months respond to gender in faces, and children as young as three years prefer to associate with others of the same gender. Interestingly, three-year-old children systematically choose to engage in activities that are practiced by other children of their own gender, relative to activities practiced by children of the opposite gender, even when the activities have been randomly assigned to the two genders.[68] These findings suggest that humans view gender as a highly important aspect not only of human biology, but of human psychology and social roles. Adults' tendency to choose professions in which their own gender is well represented may stem from predispositions that originate early in development.

This predisposition can bias our perceptions of others and of ourselves. Adults who are told that an infant is male attribute different physical properties, psychological capacities, and emotions to that infant than those who are told an infant is female, even if the labels are wrong.[69] Parents expect their male infants to have greater motoric abilities than their female infants, even on dimensions on which infants of the two sexes show objectively equal performance.[70] Orchestra boards appear to hear different qualities in the music of an auditioning applicant if they know the applicant's gender, for more female musicians are hired when auditions are gender-blind.[71] Similarly, reviewers of articles submitted to scientific journals accept fewer articles whose first authors are female when they are informed of the identity of the authors,[72] and academic faculty members look less favorably on the curriculum vitae of a prospective faculty member when the name of the candidate is female.[73] Finally, students who are reminded of their gender before taking an academic test are more likely to perform in accordance with gender stereotypes.[74] Deep-rooted beliefs in psychological gender differences can become self-fulfilling prophecies.

Gender disparities in academic professions are perpetuated by a further problem faced by those who evaluate students and select and promote faculty: It is often difficult to distinguish what is *typical* of a profession from what is *necessary* to the practice of the profession. When social factors create imbalances in the groups that practice a profession, characteristics of the dominant group may be taken, erroneously, as predictors of professional competence.[75]

In 1998, Andrew S. Winston discussed a compelling example of this mistake in a historical analysis of recommendation letters written by E. G. Boring for his Jewish students. For three decades beginning in the 1920s, Boring was the dean of American experimental psychology and director of the Psychology Laboratory at Harvard University. During that period, there were few women or Jews in American academic psychology. Indeed, Harvard did not grant PhDs to women, although it did accept Jewish students. Boring mentored a number of Jewish graduate students, and he explicitly disavowed any form of discrimination. Each student, he insisted, should be evaluated solely on the basis of his individual merits.

Nevertheless, Boring's Jewish students rarely attained prestigious academic positions, and Winston's analysis of his letters of recommendation suggested one reason why. Even when Boring described a student as the brightest and most productive of his cohort, he was apt to lament that the candidate's personal characteristics would hinder his performance. The letters written by Boring and his colleagues included numerous examples of Jewish candidates being thought to suffer from "the defects of [their] race": A candidate might be described as too "talkative," too "aggressive," or too apt to display "characteristic Jewish eagerness." These characteristics, it was suggested, would reduce the student's effectiveness as a scientist, teacher, and academic colleague.

Today, Jews are no rarity on academic faculties, and so it is easy to spot the error in Boring's thinking. In his day, successful American scientists tended to be upper-class Christian men. They were raised in a culture that encouraged them to speak softly, pause before answering questions, and let other people finish their sentences. It was natural to believe that these qualities were germane to the success of their science—that science requires the deliberative and dispassionate attitude typified by Christian temperance. These assumptions seem less reasonable today, in the multicultural world of contemporary science.

The influx of substantial numbers of women into science began later than that of Jews, and it has not progressed as far. Because most scientists and mathematicians are men, personal qualities that are stereotypically or predominantly male tend to predominate among them. Practicing scientists may tend to show characteristically male patterns of cooperation and competition, not because these patterns foster better science but because they are more common in men. Practicing mathematicians may also tend to use cognitive operations such as mental rotation more often than cognitive operations such as algebraic calculation, not because the former operations are inherently superior but because they are more prevalent in males.

Winston's analysis suggests a different interpretation of Baron-Cohen's findings relating motivational differences between males and females to differences in their orientation to science. Baron-Cohen has hypothesized that males are genetically predisposed to "systemizing" and females to "empathizing," and that these differing predispositions account for the greater success of males in science and mathematics.[76] Evidence in support of the first hypothesis is scarce, but even if it were true, we should be wary of the second. If individuals high in empathy were found to be rarer among mathematicians than among homemakers, this relationship might occur not because empathy hinders progress in mathematics, but because most practicing mathematicians are men, with a host of characteristics that are more typical of their gender. Only research, or a changing social world, will allow us to sort out the true ingredients of scientific aptitude from its incidental correlates.

In summary, two deep-rooted psychological tendencies conspire to work against the market forces that Summers viewed as antidotes to hiring inequities. First, humans are predisposed to organize the social world by gender, even in situations in which no relevant gender differences exist. Second, humans are predisposed to naturalize our current social arrangements, confusing the features that are typical of the members of a profession with those that are necessary for success in the profession. Both tendencies are apt to impede the progress of women in science.

Conclusions and Prospects

Our review of the evidence pertaining to Summers's conjectures seems to support a set of negative conclusions: There is no evidence for a male advantage in intrinsic aptitude or motivation for mathematics and science, and no reason to expect that market forces would have corrected gender inequities in these fields. The negative conclusions stem, however, from a wealth of new and exciting findings concerning the genetic basis of human cognition at its highest levels. Contrary to the claims of social construc-tionists, mathematics and science are made possible, in large part, by a set of cognitive and motivational systems that are shaped by natural selection, that emerge early in human development, and that educated adults contin-ue to call on when we reason about the world around us. Here we agree with Pinker that the mind has an intrinsic structure that shapes human thought and action.[77] This intrinsic structure goes a long distance toward explaining both our achievements and our limitations.

If the mind is not a blank slate, then surely it is reasonable to ask, as do contributors to this volume, whether genetic differences in aptitude for the sciences could account for the scarcity of women in some scientific fields. Surely the answer *could* be yes; the pressures of sexual selection could have driven male and female cognitive abilities and motivational dispositions apart, just as they have driven apart the coloration of peacocks' feathers or the courtship activities of elephant seals. Summers's conjectures are worthy of serious consideration.[78]

In this chapter, we surveyed this evidence and reject Lawrence Summers's conjectures. Research on the cognitive foundations of mathe-matical and scientific reasoning, and on the motivational patterns that bring academic success, suggests that natural selection has worked to produce convergence, not divergence, of men and women. The convergence in core cognitive and motivational patterns is deeply interesting, and it may carry important insights into the nature of human cognitive evolution. In the context of present debates and pragmatic concerns, however, it supports a simple message: Academic mathematics and science will be best enhanced if institutions work to reduce bias and barriers to women, opening the doors of the academy to the most talented scientists of both genders.[79]

Notes

1. Quinn et al. 2002.
2. Maccoby and Jacklin 1987.
3. Martin et al. 1995; Masters et al. 1979.
4. Brown et al. 2003.
5. Davison and Burke 2000.
6. Summers 2005.
7. Pinker 2002.
8. Everett 2005; Pica et al. 2004.
9. Dehaene 1997.
10. Feigenson et al. 2004; Wang and Spelke 2002.
11. Pica et al. 2004; Dehaene et al. 2006.
12. Hauser and Spelke 2004.
13. McCloskey 1983.
14. Gallistel and Gelman 1992.
15. Dehaene et al. 2006.
16. Hubbard et al. 2005; Temple and Posner 1998.
17. Lemer et al. 2003; Cappelletti et al. 2007.
18. Kim and Spelke 1999; Huntley-Fenner et al. 2002; Carey and Sarnecka 2006.
19. Gilmore et al. 2007.
20. Gilmore and Spelke 2008.
21. Shusterman et al. 2008; Geary 2006.
22. Spelke 2003.
23. Shusterman et al., 2008.
24. Gelman 1991.
25. Baillargeon 1986; Leslie 1982; Hood and Willats 1986; Wynn 1992; Berger et al. 2006.
26. Feigenson et al. 2002.
27. Maccoby and Jacklin 1974.
28. vanMarle 2004. One highly publicized experiment (Connellan et al. 2000) goes against Maccoby and Jacklin's conclusion. The authors presented newborn infants with a live, responsive person and a hand-operated moving ball; they reported that male infants showed a dramatic preference for the ball, whereas female infants showed a dramatic preference for the person. The experiment merits replication, both because its findings are at odds with a large literature on infants' visual preferences and because its method lacked controls for presenter bias. Nevertheless, note that the experiment does not test whether males show greater abilities to process objects, or even greater interest in objects, than females. It is possible that infants of both genders show equal interest in objects, and that their differing visual preferences, if replicable, stem from a gender difference in their interest in people.
29. Pinker 2002.

30. Pylyshyn and Storm 1988.

31. Scholl 2001; Carey and Xu 2001; vanMarle and Scholl 2003.

32. Green and Bavelier 2003.

33. Dehaene 1997; Dehaene et al. 2006.

34. Xu and Spelke 2000; McCrink and Wynn 2004; Temple and Posner 1998; Dehaene et al., 1999.

35. Barth et al. 2005.

36. Gilmore et al. 2007.

37. Barth et al. 2005.

38. Geary 1998.

39. Newcombe et al. 1999; O'Keefe and Burgess 1996; Chiandetti and Vallortigara 2008; Dehaene et al. 2006.

40. Cheng and Newcombe 2005.

41. Hermer and Spelke 1994.

42. Dehaene et al. 2006.

43. Hermer and Spelke 1996; Learmonth 2007.

44. Dehaene et al. 2006.

45. Grace et al. 2006.

46. Hyde 2005; Voyer et al. 1995.

47. Moore and Johnson 2008; Quinn and Liben 2008.

48. Wynn 1992; Le Corre and Carey 2007; Lipton and Spelke 2006.

49. Griffin and Case 1996.

50. Wynn 1992; Condry and Spelke 2008.

51. DeLoache 2004; Winkler-Rhoades et al. 2007.

52. Shusterman et al. 2008.

53. Ibid.

54. See also chapter 1, above.

55. Morgan 2000; Lubinski and Benbow 1992.

56. Eagly 1987; Halpern et al. 2007.

57. Bem and Allen 1974; Nisbett 2003; Dweck 2006.

58. Kagan and Fox 2006.

59. Hyde 2005; Van Vugt et al. 2007.

60. Singer et al. 2006; see also chapter 1, above.

61. Mischel et al. 1989.

62. Ayduk et al. 2000; Eigsti et al. 2006; Shoda et al. 1990.

63. Duckworth and Seligman 2005.

64. Mischel et al. 1989.

65. Silverman 2003.

66. Gallagher and Kaufman 2005; Halpern et al. 2007.

67. Kurzban et al. 2001.

68. Shutts et al., under review.

69. Condry and Condry 1976; Stern and Karraker 1989.

70. Mondschein et al. 2000. For a discussion of motoric abilities, see chapter 1, above.

71. Goldin and Rouse 2000.

72. Budden et al. 2007.

73. Steinpreis et al. 1999.

74. See chapter 5, below.

75. Valian 1998.

76. See chapter 1, above.

77. Pinker 2002.

78. Halpern et al. 2007; Ceci and Williams 2006.

79. National Academy of Sciences et al. 2007.

References

Ayduk, Ozlem, Rodolfo Mendoza-Denton, Walter Mischel, Geraldine Downey, Philip K. Peake, and Monica Rodriguez. 2000. Regulating the Interpersonal Self: Strategic Self-Regulation for Coping with Rejection Sensitivity. *Journal of Personality and Social Psychology* 79: 776–92.

Baillargeon, Renee. 1986. Representing the Existence and the Location of Hidden Objects: Object Permanence in 6- and 8-Month-Old Infants. *Cognition* 23: 21–41.

Barth, Hilary, Kristen La Mont, Jennifer Lipton, and Elizabeth S. Spelke. 2005. Abstract Number and Arithmetic in Young Children. *Proceedings of the National Academy of Sciences* 102: 14117–21.

Bem, Daryl J., and A. Allen. 1974. On Predicting Some of the People Some of the Time: The Search for Cross-Situational Consistencies in Behavior. *Psychological Review* 81: 506–20.

Berger, Andrea, Gabriel Tzur, and Michael I. Posner. 2006. Infant Brains Detect Arithmetic Errors. *Proceedings of the National Academy of Sciences of the United States of America* 103: 12649–53.

Brown, Ann J., William Swinyard, and Jennifer Ogle. 2003. Women in Academic Medicine: A Report of Focus Groups and Questionnaires, with Conjoint Analysis. *Journal of Women's Health* 12: 999–1008.

Budden, Amber E., Tom Tregenza, Lonnie W. Aarssen, Julia Koricheva, Roosa Leimu, and Christopher J. Lortie. 2007. Double-Blind Review Favours Increased Representation of Female Authors. *Trends in Ecology and Evolution* 23: 4–6.

Cappelletti, Marinella, Hilary Barth, Felipe Fregni, Elizabeth S. Spelke, and Alvaro Pascual-Leone. 2007. rTMS Over the Intraparietal Sulcus Disrupts Numerosity Processing. *Experimental Brain Research* 179: 631–42.

Carey, Susan, and Barbara W. Sarnecka. 2006. The Development of Human Conceptual Representations. In *Attention and Performance: Vol XXI. Processes of Change in Brain and Cognitive Development*, ed. M. Johnson and Y. Munakata, 73–496. Oxford: Oxford University Press.

Carey, Susan, and Fei Xu. 2001. Infants' Knowledge of Objects: Beyond Object-Files and Object Tracking. *Cognition* 80: 179–213.

Cheng, Ken, and Nora S. Newcombe. 2005. Is There a Geometric Module for Spatial Orientation? Squaring Theory and Evidence. *Psychonomic Bulletin & Review* 12: 1–23.

Chiandetti, Cinzia, and Giorgio Vallortigara. 2008. Is There an Innate Geometric Module? Effects of Experience with Angular Geometric Cues on Spatial Re-Orientation Based on the Shape of the Environment. *Animal Cognition* 11 (1): 139–46.

Condry, John, and Sandra Condry. 1976. Sex Differences: A Study of the Eye of the Beholder. *Child Development* 47: 812–19.

Condry, Kristen F., and Elizabeth S. Spelke. 2008. The Development of Language and Abstract Concepts: The Case of Natural Number. *Journal of Experimental Psychology: General* 137 (1): 22–38.

Davison, Heather K., and Michael J. Burke. 2000. Sex Discrimination in Simulated Employment Contexts: A Meta-Analytic Investigation. *Journal of Vocational Behavior* 56: 225–48.

Dehaene, Stanislas. 1997. *The Number Sense: How the Mind Creates Mathematics*. Oxford: Oxford University Press.

———, Véronique Izard, Pierre Pica, and Elizabeth S. Spelke. 2006. Core Knowledge of Geometry in an Amazonian Indigene Group. *Science* 311: 381–84.

DeLoache, Judy S. 2004. Becoming Symbol-Minded. *Trends in Cognitive Sciences* 8: 66–70.

Duckworth, Angela L., and Martin E. P. Seligman. 2005. Self-Discipline Outdoes IQ in Predicting Academic Performance of Adolescents. *Psychological Science* 16: 939–44.

Dweck, Carol S. 2006. Is Math a Gift? Beliefs that Put Females at Risk. In *Why Aren't More Women in Science? Top Researchers Debate the Evidence*, ed. S. J. Ceci and W. M. Williams. Washington, D.C.: American Psychological Association.

Eigsti, Inge-Mari, Vivian Zayas, and Walter Mischel. 2006. Predicting Cognitive Control from Preschool to Late Adolescence and Young Adulthood. *Psychological Science* 17: 478–84.

Everett, Daniel L. 2005. Cultural Constraints on Grammar and Cognition in Pirahã: Another Look at the Design Features of Human Language. *Current Anthropology* 46: 621–34.

Feigenson, Lisa, Susan Carey, and Marc Hauser. 2002. The Representations Underlying Infants' Choice of More: Object Files versus Analog Magnitudes. *Psychological Science* 13: 150–56.

Feigenson, Lisa, Stanislas Dehaene, and Elizabeth S. Spelke. 2004. Core Systems of Number. *Trends in Cognitive Science* 8: 307–14.

Gallagher, Ann M., and James C. Kaufman. 2005. *Gender Differences in Mathematics*. New York: Cambridge University Press.

Gallistel, C. R., and Rochel Gelman. 1992. Preverbal and Verbal Counting and Computation. *Cognition* 44: 43–74.

Geary, David C. 1998. *Male, Female: The Evolution of Human Sex Differences*. Washington, D.C.: American Psychological Association.

———. 2006. Development of Mathematical Understanding. In *Handbook of Child Psychology: Volume 2. Cognition, Perception, and Language*, 6th ed., ed. D. Kuhl and R. S. Siegler. New York: John Wiley and Sons.

Gelman, Rochel. 1991. Epigenetic Foundations of Knowledge Structures: Initial and Transcendent Constructions. In *The Epigenesis of Mind: Essays on Biology and Cognition*, ed. S. Carey and R. Gelman. Hillsdale, N.J.: Lawrence Erlbaum Associates.

Gilmore, Camilla K., Shannon E. McCarthy, and Elizabeth S. Spelke. 2007. Symbolic Arithmetic Knowledge without Instruction. *Nature* 447: 589–92.

Gilmore, Camilla K., and Elizabeth S. Spelke. 2008. Children's Understanding of the Relationship between Addition and Subtraction. *Cognition* 107 (3): 932–45.

Goldin, Claudia, and Cecilia Rouse. 2000. Orchestrating Impartiality: The Impact of Blind Auditions on Female Musicians. *American Economic Review* 90: 715–41.

Grace, Ariel D., Véronique Izard, Kristin Shutts, Stanislas Dehaene, Pierre Pica, and Elizabeth S. Spelke. 2006. Sensitivity to Geometry in Male and Female Children and Adults in the U.S. and in an Amazonian Indigene Group. Poster presentation at the annual meeting of the Vision Sciences Society, Sarasota, Florida.

Green, Shawn C., and Daphne Bavelier. 2003. Action Video Game Modifies Visual Selective Attention. *Nature* 423: 534–37.

Griffin, Sharon, and R. Case. 1996. Evaluating the Breadth and Depth of Training Effects When Central Conceptual Structures Are Taught. *Society for Research in Child Development Monographs* 59: 90–113.

Halpern, Diane F., Camilla P. Benbow, David C. Geary, Ruben C. Gur, Janet Shibley Hyde, and Morton Ann Gernsbacher. 2007. The Science of Sex Differences in Science and Mathematics. *Psychological Science in the Public Interest* 8: 1–51.

Hauser, Marc D., and Elizabeth S. Spelke. 2004. Evolutionary and Developmental Foundations of Human Knowledge: A Case Study of Mathematics. In *The Cognitive Neuroscience 3*, ed. M. Gazzaniga. Cambridge, Mass.: MIT Press.

Hermer, Linda, and Elizabeth S. Spelke. 1994. A Geometric Process for Spatial Reorientation in Young Children. *Nature* 370: 57–59.

———. 1996. Modularity and Development: The Case of Spatial Reorientation. *Cognition* 61: 195–232.

Hood, B., and P. Willats. 1986. Reaching in the Dark to an Object's Remembered Position: Evidence for Object Permanence in 5-Month-Old Infants. *British Journal of Developmental Psychology* 4: 57–65.

Hubbard, Edward M., Manuela Piazza, and Philippe Pinel. 2005. Interactions between Number and Space in Parietal Cortex. *Nature Reviews Neuroscience* 6: 435–48.

Huntley-Fenner, Gavin, Susan Carey, and Andrea Solimando. 2002. Objects Are Individuals But Stuff Doesn't Count: Perceived Rigidity and Cohesiveness Influence Infants' Representations of Small Numbers of Discrete Entities. *Cognition* 85: 203–21.

Hyde, Janet Shibley. 2005. The Gender Similarities Hypothesis. *American Psychologist* 60: 581–92.

Jacklin, C. N., and E. E. Maccoby. 1978. Social Behavior at Thirty-Three Months in Same-Sex and Mixed-Sex Dyads. *Child Development* 49: 557–69.

Kagan, Jerome, and Nathan A. Fox. 2006. Biology, Culture, and Temperamental Biases. In *Handbook of Child Psychology: Volume 3. Social, Emotional, and Personality Development*, 6th ed., ed. N. Eisenberg. Hoboken, N.J.: John Wiley and Sons.

Kim, K., and Elizabeth S. Spelke. 1999. Perception and Understanding of Effects of Gravity and Inertia on Object Motion. *Developmental Science* 2: 339–62.

Kurzban, Robert, John Tooby, and Leda Cosmides. 2001. Can Race be Erased? Coalitional Computation and Social Categorization. *Proceedings of the National Academy of Sciences* 98: 15387–92.

Le Corre, Mathieu, and Susan Carey. 2007. One, Two, Three, Four, Nothing More: An Investigation of the Conceptual Sources of the Verbal Counting Principles. *Cognition* 105: 395–438.

Lerner, Cathy, Stanislas Dehaene, Elizabeth S. Spelke, and Laurent Cohen. 2003. Approximate Quantities and Exact Number Words: Dissociable Systems. *Neuropsychologia* 41: 1942–58.

Leslie, Alan M. 1982. The Perception of Causality in Infants. *Perception* 11 (2): 173–86.

Lipton, Jennifer S., and Elizabeth S. Spelke. 2006. Preschool Children Master the Logic of Number Word Meanings. *Cognition* 98: B57–B66.

Lubinski, David, and Camilla P. Benbow. 1992. Gender Differences in Abilities and Preferences among the Gifted: Implications for the Math/Science Pipeline. *Current Directions in Psychological Science* 1: 61–66.

Maccoby, Eleanor E., and Carol N. Jacklin. 1974. *The Psychology of Sex Differences.* Stanford, Calif.: Stanford University Press.

———. 1987. Gender Segregation in Childhood. In *Advances in Child Development and Behavior 20*, ed. E. H. Reese, 239–87. New York, N.Y.: Academic Press.

Martin, Carol Lynn, Lisa Eisenbud, and Hilary Rose. 1995. Children's Gender-Based Reasoning About Toys. *Child Development* 66: 1453–71.

Masters, John C., Martin E. Ford, Richard Arend, Harold D. Grotevant, and Lawrence V. Clark. 1979. Modeling and Labeling as Integrated Determinants of Children's Sex-Typed Imitative Behavior. *Child Development* 50: 364–71.

McCloskey, Michael. 1983. Intuitive Physics. *Scientific American* 248: 122–30.

McCrink, Karen, and Karen Wynn. 2004. Large-Number Addition and Subtraction by 9-Month-Old Infants. *Psychological Science* 15: 776–81.

Mischel, Walter, Yuichi Shoda, and Monica L. Rodriguez. 1989. Delay of Gratification in Children. *Science* 244: 933–38.

Mondschein, Emily R., Karen E. Adolph, and Catherine S. Tamis-Lemonda. 2000. Gender Bias in Mothers' Expectations about Infant Crawling. *Journal of Experimental Child Psychology: Special Issue on Gender* 77: 304–16.

Moore, D. S., and S. P. Johnson. 2008. Mental Rotation in Human Infants: A Sex Difference. *Psychological Science* 19.

Morgan, Laurie A. 2000. Is Engineering Hostile to Women? An Analysis of Data from the 1993 National Survey of College Graduates. *American Sociological Review* 65: 316–21.

National Academy of Sciences, National Academy of Engineering, and Institute of Medicine of the National Academies. 2007. *Beyond Bias and Barriers: Fulfilling the Potential of Women in Academic Science and Engineering.* Washington, D.C.: National Academies Press.

Newcombe, Nora S., Janellen Huttenlocher, and Amy Learmonth. 1999. Infants' Coding of Location in Continuous Space. *Infant Behavior and Development* 22: 483–510.

Nisbett, Richard E. 2003. *The Geography of Thought: How Asians and Westerners Think Differently . . . and Why.* New York, N.Y.: Free Press.

O'Keefe, John, and Neil Burgess. 1996. Geometric Determinants of the Place Fields of Hippocampal Neurons. *Nature* 381: 425–28.

Pica, Pierre, Cathy Lemer, Véronique Izard, and Stanislas Dehaene. 2004. Exact and Approximate Arithmetic in an Amazonian Indigene Group. *Science* 306: 499–503.

Pinker, Steven. 2002. *The Blank Slate: The Modern Denial of Human Nature*. New York, N.Y.: Viking.

Pylyshyn, Zenon W., and Ron W. Storm. 1988. Tracking Multiple Independent Targets: Evidence for a Parallel Tracking Mechanism. *Spatial Vision* 3: 179–97.

Quinn, Paul C., Joshua Yahr, Abbie Kuhn, Alan M. Slater, and Olivier Pascalis. 2002. Representation of the Gender of Human Faces by Infants: A Preference for Female. *Perception* 31: 1109–21.

Quinn, P. C., and L. S. Liben. 2008. Sex difference in mental rotation in young infants. *Psychological Science* 11.

Scholl, Brian J. 2001. Objects and Attention: The State of the Art. *Cognition* 80: 1–46.

Shoda, Yuichi, Walter Mischel, and Philip K. Peake. 1990. Predicting Adolescent Cognitive and Self-Regulatory Competencies from Preschool Delay of Gratification: Identifying Diagnostic Conditions. *Developmental Psychology* 26: 978–86.

Shusterman, Anna B., Sang Ah Lee, and Elizabeth S. Spelke. 2008. Young Children's Spontaneous Use of Geometry in Maps. *Developmental Science* 11 (2): F1–F7.

Shutts, Kristin, Mahzarin Banaji, and Elizabeth Spelke. Under review. Social Categories Guide Young Children's Preferences for Novel Objects.

Silverman, Irwin W. 2003. Gender Differences in Delay of Gratification: A Meta-Analysis. *Sex Roles* 49: 451–63.

Singer, Tania, Ben Seymour, John P. O'Doherty, Klaas E. Stephan, Raymond J. Dolan, and Chris D. Frith. 2006. Empathetic Neural Responses Are Modulated by the Perceived Fairness of Others. *Nature* 439: 466–69.

Spelke, Elizabeth S. 2003. What Makes Us Smart? Core Knowledge and Natural Language. In *Language in Mind: Advances in the Investigation of Language and Thought*, ed. D. Gentner and S. Goldin-Meadow. Cambridge, Mass.: MIT Press.

Steinpreis, Rhea, Katie Anders, and Dawn Ritzki. 1999. The Impact of Gender on the Review of the Curriculum Vitae of Job Applicants and Tenure Candidates: A National Empirical Study. *Sex Roles* 41: 509–28.

Stern, Marilyn, and Katherine Karraker. 1989. Sex Stereotyping of Infants: A Review of Gender Labeling Studies. *Sex Roles* 20: 501–22.

Summers, Lawrence H. 2005. Remarks at NBER Conference on Diversifying the Science and Engineering Workforce. Cambridge, Mass., January 14. http://shaffner.wikispaces.com/file/view/Lawrence+Summers.pdf (accessed April 23, 2009).

Temple, Elise, and Michael I. Posner. 1998. Brain Mechanisms of Quantity Are Similar in 5-Year-Old Children and Adults. *Proceedings of the National Academy of Sciences* 95: 7836–41.

Valian, Virginia. 1998. *Why So Slow? The Advancement of Women*. Cambridge, Mass.: MIT Press.

vanMarle, K. 2004. *Infants' Understanding of Number: The Relationship Between Discrete and Continuous Quantity*. PhD dissertation. Yale University.

vanMarle, K., and Brian J. Scholl. 2003. Attentive Tracking of Objects versus Substances. *Psychological Science* 14: 498–504.

Van Vugt, Mark, David De Cremer, and Dirk P. Janssen. 2007. Gender Differences in Cooperation and Competition: The Male-Warrior Hypothesis. *Psychological Science* 18: 19–23.

Voyer, Daniel, Susan Voyer, and M. P. Bryden. 1995. Magnitude of Sex Differences in Spatial Abilities: A Meta-Analysis and Consideration of Critical Variables. *Psychological Bulletin* 117: 250–70.

Wang, Ranxiao, and Elizabeth S. Spelke. 2002. Human Spatial Representation: Insights from Animals. *Trends in Cognitive Sciences* 6 (9): 376–82.

Winkler-Rhoades, N., E. S. Spelke, and S. Carey. 2007. Of Pictures and Words. Poster presented at SRCD '07: Biennial Meeting of the Society for Research on Child Development, Boston, Mass.

Wynn, Karen. 1992. Children's Acquisition of the Number Words and the Counting System. *Cognitive Psychology* 24: 220–51.

Xu, Fei, and Elizabeth S. Spelke. 2000. Large Number Discrimination in 6-Month-Old Infants. *Cognition* 74: B1–B11.

3

A History of Structural Barriers to Women in Science: From Stone Walls to Invisible Walls

Rosalind Chait Barnett and Laura Sabattini

Where are the women in science? This oft-repeated question is still being asked today, especially with regard to the presence of female tenured professors at elite universities. Some answers echo age-old rationalizations for women's relatively poor performance in the field: History tells us that women have never made significant contributions to math or science; women are innately ill-equipped for high-level math and science careers; women who pursue demanding careers in science will jeopardize their well-being and that of their families. Some answers, however, are different, reflecting both societal changes and recent empirical data. For example, new evidence suggests that women's chances for success in science depend on the structure of the organization within which they work.[1] Women scientists are more likely to hold high-level leadership positions when working outside the hierarchical worlds of the academy and large scientific establishments.

A great deal can be learned about women's continuing lack of success in some fields of science and scientific establishments by taking a historical perspective. By situating women's experiences through time and noting how opportunities in scientific fields have shifted based on social norms, as well as on educational and organizational structures, we can then focus on steps that may help overcome the barriers that remain.

We begin this chapter with the golden age of science: the European Renaissance. What was the situation for women in the fourteenth to seventeenth centuries, when scientific endeavors were flourishing all across Western Europe? Subsequently, what was their situation in early nine-teenth-century America, when educational opportunities were expanding, especially for young males? What has been the trajectory of American women scientists' access to education, employment, and recognition throughout the nineteenth and twentieth centuries? And, finally, what is the situation for women in science today, and how might it be improved in the light of lessons from the past?

Renaissance Men and Science

Why should a serious discussion of the relative absence of women in science today start with a discussion of the Renaissance? Surely, so much has changed since then that insights from that long-ago period would appear to be of little use today. Nevertheless, some of the current rationalizations for the situation of women in science are strikingly similar to those offered in the past.

Arguably, the blossoming of science in Europe was at its peak during the fourteenth to seventeenth centuries. A fundamental transformation in scientific ideas took place in such fields as physics, mathematics, physiology, astronomy, and biology, both in institutions supporting scientific investigation and in the more widely held picture of the universe. As a result, the scientific revolution that occurred during the Renaissance is generally viewed as the foundation of modern science, and many of the scientists who lived during that incredible period are household names today, four centuries later:

- *Leonardo da Vinci* (1452–1519) was born just outside Florence, Italy. His work covered four main disciplines: painting, architecture, the elements of mechanics, and human anatomy and physiology. Over his lifetime, he produced numerous studies on such diverse subjects as nature, flying machines, geometry, mechanics, municipal construction, canals, and architecture, and he designed structures ranging from churches to fortresses. His studies contained designs for advanced weapons, including a tank and other war vehicles, various combat devices, and submarines. Da Vinci also produced detailed anatomical studies, which he recorded along with much of his other work in meticulously illustrated notebooks.

- *Nicolaus Copernicus* (1473–1543) is said to be the founder of modern astronomy. Copernicus was a Polish astronomer and mathematician and a proponent of the view that the Earth was in daily motion about its axis and in yearly motion around a stationary sun.

- *Galileo Galilei* (1564–1642) has been called the "father of modern observational astronomy," the "father of modern physics," and the "father of science." An Italian physicist, mathematician, astronomer, and philosopher, his achievements included the first systematic studies of uniformly accelerated motion, improvements to the telescope, a variety of astronomical observations, and support for Copernicanism. Galilei's experiment-based work represented a significant break from the abstract Aristotelian approach of his time.

- *Johannes Kepler* (1571–1630) was another important astronomer. He was the founder of "celestial mechanics," formulating the three laws of planetary motion and explaining how the tides were influenced by the moon. Moreover, in addition to his theories on the structure of the universe, Kepler made significant headway into the field of optics. His publication, *Stereometrica Doliorum*, provided the basis for integral calculus, and he also made important advances in geometry.

- *René Descartes* (1596–1650) is considered one of the preeminent Western philosophers of the Renaissance and beyond. During his lifetime, he was just as famous as an original physicist, physiologist, and mathematician.

- *Gottfried Leibniz* (1646–1716) was a German mathematician, philosopher, and logician who is probably best known for having invented (independently of Sir Isaac Newton, below) differential and integral calculus.

- *Sir Isaac Newton* (1642–1727), a mathematician and physicist, has been regarded for almost three hundred years as the founder of modern physical science and called the foremost English mathematician of his generation. He made major contributions to chemistry and mechanics, and, in mathematics, he laid the foundation for differential and integral calculus. Newton's work on optics and gravitation made him one of the greatest scientists the world has known.

- Remarkably, three famous mathematicians were born within three years of each other. *Pierre de Fermat* (1601–65) is thought of today as one of the greatest mathematicians and perhaps the most famous number theorist who ever lived. French scientist *Gilles Personne de Roberval* (1602–75) developed powerful methods in the early study of integration. Finally, *Bonaventura Francesco Cavalieri* (1598–1647), an Italian mathematician, made major contributions to geometry and calculus.

The fact that not one of these superstar scientists was a woman is not lost on those who point to the historical record as proof that women do not have what it takes to achieve the pinnacle of success in math and science. Is it fair to conclude that the absence of women scientists during the Renaissance is due primarily to their lack of ability, motivation, or drive? To answer this question, we need to look at what was happening to women during this historical period.

Renaissance Women and Science

Ironically, as they watched the lives and rights of their husbands, sons, and brothers expand during this period, women's lives contracted. At the height of the Renaissance, when science was flourishing, a woman generally had only four life options: to enter into a marriage, often arranged; to enter a convent; to work as a maid; or to become a prostitute. To put these options into context, consider that during the late fifteenth century, only approximately half of all eligible young Venetian women married. The prices of dowries paid to prospective husbands escalated, so to conserve their family fortunes, parents prohibited their daughters from marrying. A marriage dowry might be as much as 400 percent higher than the fee to join a convent.[2]

For daughters of the ruling class, only the first two life options were available. Young patrician girls were routinely sent away to convents to be educated and kept secure until good marriages could be arranged for them; many stayed. Indeed, many women, wealthy or not, chose to live out their lives in convents. One reason was the fear of dying in childbirth. Not surprisingly, convent life thrived throughout Europe.

A modern-day comparison gives a sense of how prevalent convent life was during the Italian Renaissance. Sixteenth-century Venice, a city with a population of 86,000, had fifty convents and about three thousand nuns.[3] The town in which one of the authors lives has a population of about 10,000—roughly one-tenth that of sixteenth-century Venice—served by one grocery store, one drugstore, and one auto-mechanics shop. This little town would have to have five convents to be proportionate to the number in Venice during the Renaissance.

What were young Renaissance women taught behind the stone walls of their convents? Generally, they studied poetry, music, embroidery, and other skills useful for managing a household. Some were taught art, others singing, but science was certainly not part of their instruction. Why would it be? Cloistered as they were, these women had no possibility of participating in the scientific life of the times.

Consider how different the story of science might have been if half the sons of the ruling classes were, as youngsters, sent away to spend their lives behind stone walls, while their sisters were free to pursue their intellectual interests.

An important premise of social psychology offers insight into some of the long-term consequences of this extreme sex segregation in science.[4] According to social role theory, when occupations and other social roles (such as family roles) are segregated by sex to the extent that science was during the Renaissance, it is human nature to infer that something inherent about each group predisposes them toward these roles—in this case men had a predisposition toward science that women clearly did not possess. Even today, we hear various explanations of how women might be inherently different from men with respect to math and science ability. Some point to differences in genetic makeup, others to hormonal differences, others to differences in brain structure, and still others to differences in motivation. Most of these "difference" advocates fail to acknowledge the role of differences in opportunities, social norms, and expectations that surely accounted, in large part, for the dearth of women scientists in the Renaissance, and in all likelihood account, at least in some part, for the underrepresentation of women in leadership positions in science today.[5]

Before moving on to discuss the persistence of these social differences into modern times, we offer two brief vignettes, one having to do with Galileo Galilei, the other with Leonardo da Vinci. Galileo, one of the most illustrious of the Renaissance scientists, had three illegitimate children: two daughters and one son. Of the three, his eldest, Virginia, was the only one who "mirrored his own brilliance, industry, sensibility, and virtue," and was, in his words, "a woman of exquisite mind."[6] Galileo deemed her unmarriageable because he had not married her mother, so when she was thirteen, he placed her and her twelve-year-old sister in a convent, where they lived out their lives in poverty and seclusion. In contrast, his son was legitimized by fiat by the grand duke of Tuscany and went off to study law at a university.

Illegitimacy also plays a central role in the second vignette. Leonardo da Vinci was the illegitimate son of a twenty-five-year-old notary and a peasant woman. Had he been born a girl, like Galileo's daughters, he would most likely have been deemed unmarriageable and sent off to spend his life behind convent walls. Bereft of congenial social norms, high expectations, and support, even this most gifted of scientists would not have been able to develop his talents.

Access to Education for American Women in Science

Convent life never took hold in non-Catholic America as it did in pre-Reformation Europe. Although there were no stone convent walls to limit women's access to science education in the United States, women found themselves kept out of higher education by the stone walls of male-only colleges and universities. During the seventeenth and eighteenth century in our history, education was largely the preserve of sons of wealthy families who were being prepared for the ministry.

An important and recurring rationalization for limiting girls' access to education was that learning would have serious negative effects on their reproductive capacity. Medical wisdom during the Victorian era (mid-nineteenth to early twentieth centuries) held that learning to read would damage women's ovaries. Echoes of such concerns can be heard today in the rationalization that managing a career and family negatively affects women's (but not men's) health, and that women don't have the physical stamina to pursue demanding work at the highest levels in math and science.[7]

Until the early nineteenth century, girls' education was, for the most part, restricted to informal learning at "Dame Schools," where instruction was offered by female teachers to very young boys and girls, usually in the teachers' homes. Young girls were taught basic reading and writing, embroidery, and other "feminine" skills. Beginning in the mid-1820s, however, the United States rapidly became the world leader in public and private education for girls and women. Although they continued to be excluded from the science education offered by the male-only colleges, women could take advantage of a proliferation of popular books and textbooks written for them by both men and women on such subjects as botany, chemistry, and geology.[8]

The sales figures for these books reflected an enormous audience hungry for them. *Conversations on Chemistry* (Marcet 1806) went through more than fifteen editions in the United States before 1860; *Familiar Lectures on Botany* (Phelps 1829) went through at least seventeen editions and sold over 275,000 copies by 1872; *Introduction to Botany, in a Series of Familiar Letters* (Wakefield 1796) had at least nine English editions by 1841. Apparently, women's interest in pursuing science, although severely hampered by the restrictive educational and social norms of the times, was not deterred.[9]

Over the following decades, women's opportunities for access to education in science increased, but not as much as one would have imagined, given their interest and what seemed like promising opportunities. One of the most striking examples of the failure of women to increase their representation in science education occurred after World War II. Following the passage of the G.I. Bill, veterans were granted preferential treatment at all educational institutions, including women's colleges, and at all levels of the education system. As a result, the number of males earning PhDs in the post-war period skyrocketed, while the figures for women did not appreciably change. Indeed, women were often rejected from PhD programs to accommodate male veterans who, although deserving of special consideration, may have been less able and less qualified.[10]

Employment for Women Scientists

In the nineteenth century, women's access to education and opportunities for employment in science were greatly enhanced by the advent of the women's colleges, as is well documented in the encyclopedic historical work of Margaret Rossiter (1982, 1995), upon which we rely for much of the material in this section.

The mid-1800s saw the opening in the United States and elsewhere of several such colleges, many of which offered extensive courses in the sciences, especially astronomy, botany, and chemistry.[11] These schools prospered, and women graduates were often hired on as teachers of the next generation. The limited opportunities for employment in science came, however, at a considerable personal price. All women college faculty had to be single; if they decided to marry, they had to resign—a practice that in some parts of the Western world continued well into the twentieth century.[12] Once again we see a clear manifestation of the belief that women could not handle both professional and family responsibilities. In addition, the women's colleges had very heavy teaching loads, precluding faculty from conducting publishable research. Nevertheless, these were the best jobs to which women could aspire in science and, according to Rossiter, women were happy to have them.[13]

The employment situation for women in science took a turn for the worse in the early twentieth century as the women's colleges began to focus on increasing their prestige. One important step was to require faculty members to have PhD degrees; another was to recruit male faculty. To attract male applicants the colleges had to offer strong incentives, including no restriction on marriage; indeed, young married men with families were preferred candidates. The men were also offered reduced teaching loads, higher salaries, college-funded support for their research, and allowances for family living expenses.

This two-tier hiring system produced several negative consequences for women faculty. First, few women scientists at the time had doctoral degrees, because most European and U.S. universities refused to allow women to matriculate into their graduate schools. Even after some graduate schools opened their doors to women, beginning in the 1890s, not all were so welcoming. Indeed, Princeton, New York University, and Harvard did not grant women PhDs until the 1960s. Second, male faculty began publishing research, whereas women science faculty, still burdened by lack of support and heavy teaching schedules, did not. Finally, women's colleges ceased to be the primary employer for women scientists; their mission became to educate women scientists, not hire them.

Eventually some barriers fell, but others arose. For example, as the marriage ban was phased out (as was the pregnancy ban), anti-nepotism rules took hold. A woman scientist married to a male scientist would not be hired by the same department and was often refused employment at the same college or university. The result was that many highly educated, married women scientists could not find employment, especially in small cities and towns with only one academic institution.

Anti-nepotism rules were succeeded by the "two-body problem" in the hiring of women for academic positions in science, and it remains a barrier to their advancement in the field today. Simply stated, when one partner in an academic couple is offered a position at a university, the other spouse (typically the woman) also needs to find appropriate employment. Rarely is it possible to place two high-level academics at the same university. More often than not in such situations, the female is offered no position or a position at a lower level than the one she left. Thus, relocation is often beneficial to the male in the couple and detrimental to the career advancement of the "trailing spouse."

In these myriad ways, the academy has not been a hospitable employer for women in science, having erected too many structural barriers—too many unscalable walls—some of which remain today.[14] Many who could leave the academy did so. Indeed, the best-known women scientists of the twentieth century were not members of any academic faculty. They included Margaret Mead, who was never offered a tenured position and worked for most of her professional life at the American Museum of Natural History; Rachel Carson, perhaps the most influential woman scientist of the century, who supported herself by writing books, including the international bestseller, *Silent Spring*; Barbara McClintock, a cytogeneticist, who was denied tenure at the University of Missouri and went to work at Cold Spring Harbor Laboratory, where she won the Nobel Prize in Physiology and Medicine; and Dian Fossey, who completed her groundbreaking fieldwork with gorillas in Rwanda before getting her doctoral degree.

Data collected over the past twenty years have shown that, despite increasing numbers of women with science and engineering degrees, gender representation in the academy remains uneven, with men still outnumbering women at all faculty levels.[15] Recent research suggests that, even today, women scientists' chances for advancement in the academy and other hierarchically organized scientific establishments are relatively poor. Women employed full-time in academia are less likely than men to be tenured and, on average, they earn less than their male counterparts.[16] Furthermore, women are less likely than men to be employed at the highest-tiered academic institutions.[17]

According to a comparative study by Laurel Smith-Doerr, biotechnology firms with flatter, more interconnected forms of organization are better workplaces for female scientists. Smith-Doerr found that for male scientists, the odds of achieving supervisory rank were unrelated to the structure of the organization in which they worked. In contrast, for female scientists, organizational structure made an enormous difference: Females in biotech firms were eight times more likely than their male counterparts in hierarchical settings to have supervisory positions.[18] Similarly, Lois Joy found few differences in the advancement opportunities of women and men scientists employed in the industrial and health-care sectors in the United States.[19]

How can we account for these differences between academia and industry? One explanation is that networked organizations rely on partnerships to succeed and are more flexible and transparent than hierarchical organizations. As a result, advancement is based on input from a wide range of people rather than a few, as is typical in hierarchical organizations, providing fewer chances for sexism to thrive and more opportunities to create work environments that are welcoming to women.[20] Joy also argues that the criteria for promotion in academia are especially susceptible to the influence of subjective evaluations (and hence to gender bias), whereas industry pressures to achieve specific business results (e.g., the creation of new products) shift the focus onto more explicit and somewhat less subjective standards, such as the ability to bring new discoveries and successful products to the market.[21]

Not surprisingly, women scientists are increasingly choosing to work in nonacademic settings. In 2002, only 42 percent of women PhDs in science and engineering worked at universities and four-year colleges.[22] In fact, the number of U.S. life scientists, men and women, working outside the academy grew from 83,000 in 1980 to 181,000 in 2000.[23] In 2006, according to data from the National Science Foundation, the majority of scientists and engineers worked in the business/industry sector (69.4 percent), followed by educational institutions (18.8 percent) and government (11.8 percent).[24]

In spite of the greater growth in employment and some of the most remarkable developments in medicine having come from the private sector, analyses of gender relations in science are largely based on studies of the academy. It seems as if women are rejecting the persistent idea that any PhD worth her salt obtains a university position, with other options considered second-best.[25] Thus, to understand how women scientists are doing professionally, it behooves researchers to look beyond the academy.

Advancement Opportunities

In both academic and business settings, women continue to be underrepresented in key decision-making, administrative, and management positions in the sciences. As recently as 2000, women held only 4 percent of all department-head positions at the top fifty university chemistry programs in the United States, and there were no women department heads in the top fifty chemical engineering or top fifty physics programs.[26] In nonacademic settings in 2006, women constituted only 28 percent of all management or administration professionals employed in engineering[27] and held only 17 percent of senior management positions at life sciences companies, where no increase was recorded between 2002 and 2007.[28]

Research looking at gender representation among doctoral-level scientists employed in the technology sector has also shown that women were 50 percent less likely than their male counterparts to be employed in science and engineering (S&E) jobs and, even when they were, they received 20 percent less in pay.[29] 2007 *Catalyst Census of Women Board Directors of the Fortune 500* data show that, in 2006, women in the professional, scientific, and technical industries sat on only 10 percent of the boards of directors and held only 13–14 percent of the highest executive positions among the larger U.S. manufacturing and technology firms; women were similarly underrepresented on the boards and in corporate officer positions in the largest utilities and information technology and health-care institutions.[30] These numbers are especially low considering that, in 2007, women represented more than 40 percent of medical scientists and biologists and more than 40 percent of materials scientists and chemists.[31]

Despite these disparities, the number receiving degrees in science and engineering has increased over the past few decades. The National Academy of Sciences reported that, in 2006, Massachusetts Institute of Technology (MIT) science and engineering undergraduates included, respectively, 51 percent and 35 percent women. Joy has also calculated that women's share of doctoral, master's, and bachelor's degrees in science and engineering has tripled, on average, in the past decades. For example, between 1966 and 2004, the percentage of women completing a PhD went from 5.8 to 30.3; similarly, the proportion of women completing an S&E master's program went from 9.6 percent to 32 percent.[32] It seems, hence, important that we understand the factors that keep women from advancing in the field, or that might dissuade them from pursuing careers in the sciences, even after obtaining a degree.

Receiving Appropriate Recognition for Scientific Accomplishments

The historical record clearly shows that women scientists, despite their outstanding contributions, have rarely received commensurate recognition. Among the female "almost" Nobel Laureates are the following:

- Biochemist *Viola Graham*, who helped James Sumner of Cornell University synthesize urease, for which he shared the 1946 Nobel Prize in Chemistry with two other men.

- Physicist *C. S. Wu* of Columbia University, who performed the crucial experiments proving the theory that won her colleagues, Lee and Yang, the 1957 Nobel Prize in Physics.

- Geneticist *Esther Lederberg*, who helped her then-husband Joshua with the microbial research that won him and two other men the Nobel Prize in 1958.

- Biochemist *Rosalind Franklin*, who was responsible for much of the research and discovery work that led to an understanding of the structure of DNA, for which Watson, Crick, and Wilkins received a Nobel Prize in 1962.

- Biochemist *Ruth Hubbard*, who had done important work on the chemistry of vision before marrying her husband, George Wald, in 1958. Wald won the 1967 Nobel for work in the same area. Many assumed that they had always collaborated, and that he deserved most of the credit for her earlier independent work as well as their joint efforts. According to Rossiter, "Once she married a scientist of greater reputation, a woman's own independent work would all too easily be dismissed as merely a small part of his."[33]

- *Marguerite Vogt*, a molecular biologist and the colleague and close collaborator for twenty years of Renato Dulbecco in research on DNA tumor viruses and cell growth. Dulbecco shared the Nobel Prize with two other men in 1975.

- Economist *Anna Schwartz*, who coauthored several books with Milton Friedman and worked for decades on the detailed economic data that formed the basis for work that won him (alone) the economics prize in 1976.

- Crystallographer *Isabella Karle*, who worked for a lifetime with her husband, also a crystallographer, while he shared the chemistry prize in 1985 with two other men.

Although women scientists today are more likely than in the past to receive the recognition they deserve, their situation is far from equitable, especially in the academy. The problem of gender bias received national attention when it was discovered that senior faculty women at MIT were not regarded nearly as well as their men counterparts. In comparison with men senior faculty, women faculty with equal professional accomplishments were less likely to receive equitable salaries, laboratory space, research grants and awards, support, and other resources. Moreover, they were increasingly marginalized by their departments and, over time, excluded from playing significant roles.[34] The most likely explanation for their treatment was the deeply entrenched (albeit unconscious) gender bias of academic science, including work environments inhospitable to women, gender stereotypes, bias in performance evaluations, and other structural barriers.[35] Although the details differ, these obstacles echo well-documented and deeply entrenched gender bias of academic science dating back to the nineteenth century.

Despite these invisible walls, women scientists have made important inroads into the top ranks of leadership at the university level:

- *Shirley Ann Jackson,* a theoretical physicist, is president of Rensselaer Polytechnic Institute and the first African-American woman to receive a doctorate from the Massachusetts Institute of Technology.

- *Shirley M. Tilghman*, a molecular biologist, is president of Princeton University.

- *Mary Sue Coleman*, a biochemist, is president of the University of Michigan.

- As provost and senior vice president of academic affairs, *M. R. C. Greenwood*, an expert in genetics and nutrition, holds the second-highest post in the University of California system.

- *Kim Bottomly*, an immunobiologist, is president of Wellesley College.

Moreover, outside the academy, a growing number of women scientists now head major scientific and engineering institutes. For example, Claire Fraser, president and director of the Institute for Genomic Research, leads research teams that have sequenced the genomes of several microbial organisms that cause anthrax, Lyme disease, syphilis, tuberculosis, cholera, meningitis, pneumonia, and ulcers, among other diseases. Judith Rodin, a research psychologist, is president of the Rockefeller Foundation. In addition, as of 2006, four of twenty-seven institute directors at the U.S. National Institutes of Health were women.

This overview is not exhaustive; it is merely meant to illustrate that many organizations other than universities provide opportunities for women scientists to achieve prominent positions of leadership. While it is certainly encouraging that some women scientists are breaking through the "glass ceiling," research suggests that barriers still block the advancement of many others. One of the most telling studies was conducted in Sweden, arguably the country with the best record on gender equity.

The researchers, Wennerås and Wold,[36] were curious about the poor success rate among women applicants for prestigious fellowships offered by the Swedish Medical Research Council. They noted that during the 1990s, female scientists applying for these much-sought-after fellowships had been less than half as successful as male applicants. They wondered whether gender bias might be affecting the selection process. To find out, they obtained the evaluations by the teams of peer reviewers who judged the applications. Applicants had to submit their curriculum vitae and proposals for the work they intended to do if their applications were successful. The reviewers rated each application on three subjective criteria: relevance of the research proposal, scientific competence, and the quality of the proposed methodology.

An examination of the evaluations indicated that the peer reviewers rated the women applicants lower on all three criteria, but especially on scientific competence, typically reflecting the number and quality of their scientific publications. The inference was that the female applicants were less productive than the male applicants. But were they? To answer that question, Wennerås and Wold scored each of the applications on six objective criteria, comprising the total number of scientific publications, the number of publications on which the applicant was the first author, and four indicators of the impact of the applicant's publications. "Impact" referred to the "number of times the average paper in a particular journal [was] cited during one year: articles in high impact journals are cited more often and therefore have greater impact than articles in low impact journals."[37]

When the researchers compared the scores on the six objective criteria to the subjective scores the applicants received from the peer reviewers, they found clear evidence of egregious gender bias: "The most productive group of female applicants," they wrote, "containing those with 100 total impact points or more, was the only group of women judged to be as competent as men, although only as competent as the least productive group of male applicants (the one whose members had fewer than 20 total impact points)."[38] Further analyses revealed that for a female applicant to be awarded the same competence score as a male colleague, she would have to produce approximately three extra papers in high-impact journals such as *Nature* or *Science*, or twenty extra papers in excellent specialist journals such as *Atherosclerosis, Gut, Infection and Immunity, Neuroscience,* or *Radiology.* Thus, a female applicant had to be *two and a half times* more productive than the average male applicant to receive the same competence score.

This extraordinary study provides direct evidence that the peer-review system is subject to sex bias. Wennerås and Wold illuminated what would otherwise have been an "invisible wall" impeding the advancement of women in science. Such sex bias had demonstrably negative effects on the female applicants' odds of obtaining important research fellowships and thus increasing their chances for future advancement and recognition. Recall that these results were obtained in Sweden, where gender bias is reputedly low. Not surprisingly, scientists, even in Sweden, are no less immune than other human beings to the effects of prejudice.

Organizational and structural approaches are the best way to address invisible barriers, such as those inherent to the peer-review system, and stereotyped perceptions of women in the sciences. Focusing on the structural constraints within workplace and academic environments tackles the problem at its root and discourages the view that women's individual interests and cognitive abilities are to blame for gender differences in the sciences.[39] We turn next to structural solutions that may be conducive to the advancement of women in the sciences.

Addressing Gender Inequality in Science: A Structural Approach

The underrepresentation of women in the sciences—both in academia and in industry—has important consequences at all levels, for women, for educational institutions, and for businesses. Take the dearth of women in decision-making roles that we discussed earlier. To the extent to which managers and administrators oversee institutional (financial and human capital) as well as strategic resources, they also influence the expansion of scientific knowledge and production. For nonacademic industries, this influence includes making decisions about the development and production of market goods.[40] It follows that not only do women's perspectives and contributions in the sciences remain underappreciated, but also gender inequality at the very top can have significant and possibly negative consequences for the growth of scientific knowledge, productivity, and profitability.

Studies suggest that research and development teams composed of both men and women are better equipped than less diverse teams to create innovative products, and they even enhance individual performance.[41] According to a recent benchmarking study by the Healthcare Business Women's Association, companies in the life sciences industry and marketplace that proactively attract and develop women's talent have a considerable advantage over their competitors.[42] All in all, as is the case for other types of diversity, gender diversity is important to keep in touch with consumer markets, help attract talented scientists from a variety of backgrounds, and enhance productivity and creativity.[43]

Implementing strategies that directly address structural barriers is a good first step to promoting gender equality. Although programs and practices to increase the representation of women in the sciences have been around for some time, their scope and objectives have evolved as research has started to uncover the root causes of inequity, while challenging some of the previous assumptions as to why more women aren't entering these fields. Gilbert, for example, notes that early practices focused mainly on the educational pipeline and on increasing women's interest in science. These strategies included providing girls and women with access to important information about science-related careers, as well as to positive women role models. Other programs have sought to build background knowledge and skills that women (presumably) lack.[44] While such educational programs that support women can, indeed, be beneficial to individuals and provide them with important tools and information, they do not address the root causes of women's underrepresentation in science. More systematic approaches should address the invisible walls and barriers that women still have to face even after they decide to pursue a career in the field.

The EDGE (Empowerment, Diversity, Growth and Excellence) in Leadership study[45] provides organizations with a number of recommendations for addressing structural barriers to women's advancement in the pharmaceutical and biotechnology industries. Many of these recommendations can be applied to other male-dominated fields and contexts in the workplace. Specifically, the EDGE study identifies six practices that seem essential to creating a successful initiative to advance women in the industry:

- Senior leadership support for corporate changes

- Equitable performance evaluation processes

- Measures and accountability that focus on specific behaviors and drive results

- Recruitment practices that support equal representation of women

- Advancement programs for high-potential women

- Career and work flexibility to retain talent

To gain the most benefit in terms of women's advancement, the EDGE study suggests that employers should implement these six practices systematically, because relying on only one approach at a time might not work. Companies and educational institutions alike can support the practices in a number of ways, such as by instituting mentoring and career-developmental programs, setting clear objectives for advancing women, and increasing accountability and transparency in talent management processes, as well as measuring progress along the way.

Also important to overcoming structural barriers is providing individual female and male employees with the tools and infrastructures necessary to their success. Examples include high-potential employee programs and a set of clear expectations regarding advancement opportunities. Finally, workplace flexibility (including career-path flexibility) and inclusiveness are essential to retaining top talent, as they provide the opportunity for women and men from different backgrounds to excel.[46]

Structural approaches can also be helpful in academic institutions. In a recent analysis of faculty hiring processes at science, mathematics, and engineering departments in a university setting and over a period of five years, for example, sociologists Glass and Minnotte identified a number of factors associated with the successful hiring of women in science faculty jobs. The most successful searches had at least one woman on the search committee, had advertised the position in publications targeted at women scientists, and had comprised a larger pool of women applicants.[47] The research found that, despite its apparent willingness to hire more women, the university worked from a largely male-dominated pool and sometimes was at a loss in terms of how to appeal to more women to apply for the position. Faculty searches that posted in publications focused on women in science attracted more female applicants into the pool than other searches. The authors argue that this approach may help compensate for men's "information advantage" when it comes to learning about openings through informal networks and sources.

To break down the invisible barriers discussed in this chapter, it is also important to continue raising awareness about gender inequities in science within academic and educational institutions. Economist Ginther, for example, recommends that colleges and universities regularly evaluate the status of women in their science departments, as some of the larger institutions are already doing.[48] Raising awareness among faculty, administrators, and students can also help draw attention to gender differences outside academia, including differences in salary, promotion, and attrition between academia and industry. An integrated approach that seeks to overcome existing barriers in both contexts is most likely to succeed.

Important steps toward creating institutional support for advancing women in science are already being taken. These steps occur at different points along girls' and women's educational and professional pathways, and may include mentoring programs, proactive recruitment strategies, knowledge sharing, and scholarship programs. The following are some creative and interesting examples:

- The Junior Engineering Technical Society (JETS) leadership fund provides $5,000 academic scholarships to students who intend to take engineering courses in college with an eye toward careers in the power generation industry. In 2008, the four awards were split evenly by gender.[49]

- The RAISE Project seeks to increase the status and visibility of professional women through enhanced recognition of their achievements in science, medicine, and engineering. The program includes more than a thousand awards and features a database with information for applicants from different disciplines and at different stages in their careers.[50]

- The National Science Foundation's ADVANCE grants have the stated goal of increasing the participation and advancement of women in academic science and engineering careers. The program also supports institutional efforts and other women's leadership initiatives that seek to make academia more hospitable for women scientists.[51] ADVANCE's Institutional Transformation Awards are granted to academic institutions that have implemented programs to promote the advancement of women scientists and engineers.[52]

- The Women's Initiative at MIT seeks to encourage young women to pursue careers in engineering through a number of mentoring programs. The initiative is organized so that every year, eleven female engineering students from MIT visit thousands of high school students in eleven different districts throughout the United States. By providing role models and information on engineering degrees, the initiative seeks to motivate girls to take the most challenging math and science courses offered in their high schools.[53]

Our hope is that a combination of efforts such as those described in this chapter will have the much-desired effect of tearing down the invisible walls that now impede women's progress in the sciences.

Notes

1. Smith-Doerr 2004.

2. Laven 2003.

3. Ibid.

4. For a recent overview of social role theory, see Eagly et al. 2000.

5. Some arguments put forth today presume that psychological differences between women and men explain unequal gender representation in a number of contexts and roles, including leadership roles (holding, for example, that women are less ambitious than men). In an empirical article, however, psychologist Janet Hyde shows strong support for a "gender similarities hypothesis," indicating that women and men are more similar than different on a large number of psychological variables; see Hyde 2005. For a specific discussion of similarities between boys' and girls' math and science abilities, also see Hyde and Linn (2006, 599–600), as well as chapter 2, above.

6. Sobel 2000, 4.

7. Although beyond the scope of this chapter, it seems important to note that rationalizations purported to "protect" women or advocate for their health and well-being by regulating their commitments actually limit women's opportunities and constitute a form of discrimination. For a discussion of this type of "benevolent" sexism, see Glick and Fiske 2001.

8. Rossiter 1982, 1–15.

9. Ibid., 3.

10. Rossiter 1995, 28–33.

11. Rossiter 1982, 1–15.

12. Italy, for example, had to pass a law in 1963 that challenged the then common practice of employers' dismissing women upon marriage so they could avoid covering maternity leave. Pojmann 2005, 67.

13. Rossiter 1982, 51–72.

14. Biernat et al. 2004; Sabattini and Crosby 2008.

15. Joy 2008.

16. Ginther 2001; Joy 2008; National Science Foundation 2004.

17. Joy 2008.

18. Smith-Doerr 2004.

19. Joy 2008. As we discuss later, however, Joy also found that gender gaps in managers' pay persisted despite the fewer barriers to promotion opportunities.

20. Glick and Fiske 2007.

21. Joy 2008.

22. National Science Foundation 2002.

23. National Science Foundation 2002; National Science Foundation, Division of Science Resources Statistics 2001, table 19.

24. Stine and Matthews 2008, 8.

25. Smith-Doerr 2004, 59.

26. AWIS 2008.

27. National Science Foundation, Division of Science Resources Statistics 2006, table H-33.

28. Pettersson et al. 2007.

29. Graham and Smith 2005; Joy 2008.

30. Catalyst 2007b; U.S. Department of Labor, Bureau of Labor Statistics, and U.S. Bureau of the Census 2008, table 11; National Center for Women and Information Technology (NCWIT) 2007.

31. Specifically, in 2007, percentages of women in the labor force included 42.6 percent of biologists, 49.1 percent of medical scientists, and 40.8 percent of chemists and material scientists. Women's representation was lower among computer and mathematical occupations, at 25.6 percent. U.S. Department of Labor, Bureau of Labor Statistics, and U.S. Bureau of the Census 2008, table 11.

32. Joy 2008. See also National Academy of Sciences et al. 2007 on gender differences in promotion and pay.

33. Rossiter 1982, 331.

34. Hopkins et al. 2002. For another view of this study, see chapter 4, below.

35. These barriers are also well documented in National Academy of Sciences et al. 2007.

36. Wennerås and Wold 1997. For critical commentaries on this study, see chapters 4 and 9, below.

37. Ibid., 342.

38. Ibid.

39. Catalyst 2007a.

40. Joy 2008.

41. Ibid.; Turner 2006.

42. Pettersson et al. 2007.

43. Joy 2008; Turner 2006.

44. Gilbert 2001.

45. Pettersson et al. 2007.

46. D'Annolfo Levey et al. 2008.

47. Jaschik 2008.

48. Ginther 2004, 6–10.

49. Wagman 2008. For more information see also http://www.jets.org/.

50. RAISE Project 2007–8.

51. Ginther 2004.

52. ADVANCE 2008.

53. Massachusetts Institute of Technology 2008.

References

ADVANCE. 2008. ADVANCE portal website homepage. http://research.cs.vt.edu/advance/index.htm (accessed January 9, 2009).

AWIS. 2008. Table: Utilization of Chemistry, Chemical Engineering, Physics, and Biological Sciences PhD Recipients in Faculties of Those "Top 50" Departments. http://www.serve.com/awis/statistics/utilizationPHYS.pdf (accessed January 9, 2009).

Biernat, Monica, Faye J. Crosby, and Joan Williams, eds. 2004. The Maternal Wall: Research and Policy Perspectives on Discrimination against Mothers. Journal of Social Issues, vol. 60, no. 4. Boston, Mass.: Wiley-Blackwell.

Catalyst. 2007a. The Double-Bind Dilemma for Women in Leadership. New York, N.Y.: Catalyst.

Catalyst. 2007b. 2007 Catalyst Census of Women Board Directors of the Fortune 500. New York, N.Y.: Catalyst.

D'Annolfo Levey, Lisa, Meryle Mahrer Kaplan, and Aimee Horowitz. 2008. Making Change-Beyond Flexibility: Work-Life Effectiveness as an Organizational Tool for High Performance. New York, N.Y.: Catalyst.

Eagly, A. H., W. Wood, and A. B. Diekman. 2000. Social Role Theory of Sex Differences and Similarities: A Current Appraisal. In The Developmental Social Psychology of Gender, ed. T. Eckes and H. M. Trautner, 123–74. Mahwah, N.J.: Erlbaum.

Gilbert, J. 2001. Science and Its "Other": Looking Underneath "Woman" and "Science" for New Directions in Research on Gender and Science Education. Gender and Education 13 (3): 291–305.

Ginther, Donna K. 2001. Does Science Discriminate Against Women? Evidence from Academia, 1973–1997. Federal Reserve Bank of Atlanta Working Paper No. 2001-2. February. http://ssrn.com/abstract=262438 (accessed January 9, 2009).

———. 2004. Why Women Earn Less: Economic Explanations for the Gender Salary Gap in Science. AWIS Magazine 33 (1): 6–10.

Glick, Peter, and Susan T. Fiske. 2001. Ambivalent Sexism. In Advances in Experimental Social Psychology, ed. Mark P. Zanna 33: 115–88. Thousand Oaks, Calif.: Academic Press.

———. 2007. Sex Discrimination: The Psychological Approach. In Sex Discrimination in the Workplace, ed. F. J. Crosby, M. S. Stockdale, and S. Ann Ropp. Malden, Mass.: Blackwell.

Graham, J., and S. Smith. 2005. Gender Differences in Employment and Earnings in Science and Engineering in the US. Economics of Education Review 24: 341–54.

Hopkins, Nancy, Lotte Bailyn, Lorna Gibson, and Evelynn Hammonds. 2002. The Status of Women Faculty at MIT: An Overview of Reports from the Schools of Architecture and Planning; Engineering; Humanities; Arts, and Social Sciences; and the Sloan School of Management. MIT Faculty Newsletter 14 (4): 1–28.

Hyde, Janet S. 2005. The Gender Similarities Hypothesis. American Psychologist 60 (6): 581–92.

Hyde, Janet, and Marcia C. Linn. 2006. Gender Similarities in Mathematics and Science. *Science* 314 (5799): 599–600.

Jaschik, Scott. 2008. Keys to Hiring Women in Science. *Inside Higher Ed News.* August 5. http://www.insidehighered.com/news/2008/08/05/women (accessed January 9, 2009).

Joy, Lois. 2008. *Women in Health Care and Bioscience Leadership State of the Knowledge Report: Bioscience, Academic Medicine, and Nursing—Glass Ceilings or Sticky Floors?* New York, N.Y.: Catalyst.

Laven, Mary. 2003. *Virgins of Venice: Enclosed Lives and Broken Vows in the Renaissance Convent.* N.Y.: Penguin Books.

Massachusetts Institute of Technology. 2008. Women's Initiative. http://web.mit.edu/wi/sponsors (accessed January 9, 2009).

National Academy of Sciences, National Academy of Engineering, and Institute of Medicine of the National Academies. 2007. *Beyond Bias and Barriers: Fulfilling the Potential of Women in Academic Science and Engineering.* Washington, D.C.: National Academies Press.

National Center for Women and Information Technology (NCWIT) 2007. *Fact Sheet.* http://www.ncwit.org/about.factsheet.html (accessed January 9, 2009)

National Science Foundation. 2002. *Women, Minorities, and Persons with Disabilities in Science and Engineering: 2000.* Arlington, Va.: National Science Foundation.

———. 2004. Gender Differences in the Academic Careers of Scientists and Engineers. http://www.nsf.gov/statistics/nsf04323/pdf/nsf04323.pdf (accessed January 9, 2009).

———. Division of Science Resources Statistics. 2001. *2001 Survey of Doctorate Recipients.* Table 19. Employed Doctoral Scientists and Engineers in Universities and 4-Year Colleges, by Broad Field of Doctorate, Sex, Faculty Rank, and Years since Doctorate: 2001. http://www.serve.com/awis/statistics/Faculty_rank_yrs_since_doctorate_sex_2001.pdf (accessed January 12, 2009).

———. 2006. *Women, Minorities, and Persons with Disabilities in Science and Engineering.* Table H-33. Primary or Secondary Work Activity of Scientists and Engineers Employed in Business or Industry, by Age, Sex, Race/Ethnicity, and Disability Status: 2006. http://nsf.gov/statistics/wmpd/pdf/tabh-33.pdf (accessed January 12, 2009).

Pettersson, Anna K., Barbara Pritchard, and HBA EDGE in Leadership Study Team. 2007. The Progress of Women Executives in Pharmaceuticals and Biotechnology: A Leadership Benchmarking Study. Healthcare Business Women's Association and Booz, Allen, Hamilton, Inc. http://www.hbanet.org/Research/EDGE.aspx (accessed January 9, 2009).

Pojmann, Wendy. 2005. Emancipation or Liberation? Women's Associations and the Italian Movement. *Historian* 67 (1): 73–96.

RAISE Project. 2007–8. The RAISE Project. http://www.raiseproject.org (accessed January 9, 2009).

Rossiter, Margaret W. 1982. *Women Scientists in America: Struggles and Strategies to 1940.* Baltimore, Md.: Johns Hopkins University Press.

————. 1995. *Women Scientists in America: Before Affirmative Action 1940–1972.* Baltimore, Md.: Johns Hopkins University Press.

Sabattini, L., and F. J. Crosby. 2008. Work Ceilings and Walls: Work-Life and Family-Friendly Policies. In *The Glass Ceiling in the 21st Century: Understanding Barriers to Gender Equality*, ed. M. Barreto, M. Ryan, and M. Schmitt. Washington, D.C.: American Psychological Association.

Smith-Doerr, Laurel. 2004. *Women's Work: Gender Equality vs. Hierarchy in the Life Sciences.* London: Rienner.

Sobel, Dava. 2000. *Galileo's Daughter.* New York: Penguin Books.

Stine, Deborah D., and C. M. Matthews. 2008. *The U.S. Science and Technology Workforce.* Congressional Research Service. Report RL34539. June 20. http://wikileaks.org/leak/crs/RL34539.pdf (accessed April 23, 2009).

Turner, Laure. 2006. Gender Diversity: A Business Case? In *Women in Science and Technology: The Business Perspective.* European Commission. Brussels. http://ec.europa.eu/research/science-society/pdf/wist_report_final_en.pdf (accessed January 9, 2009).

U.S. Department of Labor, Bureau of Labor Statistics, and U.S. Bureau of the Census. 2008. *Current Population Survey.* Annual Averages. Table 11: Employed Persons by Detailed Occupation, Sex, Race, and Hispanic or Latino Ethnicity 2007. http://www.bls.gov/cps/cpsaat11.pdf (accessed January 12, 2009).

Wagman, David. 2008. Four Reasons for Optimism. *Power Engineering.* June. http://pepei.pennnet.com/display_article/331863/6/ARTCL/none/none/1/Four-Reasons-for-Optimism/ (accessed January 9, 2009).

Wennerås, Christine, and Agnes Wold. 1997. Nepotism and Sexism in Peer-Review. *Nature* 387 (22): 341–43.

4

Sex, Science, and the Economy

Christina Hoff Sommers

Math 55 is advertised in the Harvard University course catalog as "probably the most difficult undergraduate math class in the country."[1] It is legendary among high school math prodigies, who hear terrifying stories about it in their computer camps and at the Math Olympiads. Some go to Harvard just to have the opportunity to enroll in it. Its formal title is "Honors Advanced Calculus and Linear Algebra," but it is also known as "math boot camp," and the class compared to "a cult."[2] The two-semester freshman course meets for three hours a week, but, the catalog says, homework takes between twenty-four and sixty hours a week.[3]

Math 55 classes do not look like America. Each year as many as fifty students sign up, but at least half drop out within a few weeks. As one former student told the *Harvard Crimson* newspaper in 2006, "We had 51 students the first day, 31 students the second day, 24 for the next four days, 23 for two more weeks, and then 21 for the rest of the first semester."[4] Said another student, "I guess you can say it's an episode of 'Survivor' with people voting themselves off." The final class roster, according to the *Harvard Crimson*: "45 percent Jewish, 18 percent Asian, 100 percent male."[5]

Why do women avoid classes like Math 55? Why, in fact, are so few women in the higher echelons of academic math and in the physical sciences?

Women now earn 57 percent of bachelor's degrees and 59 percent of master's degrees overall.[6] According to the Survey of Earned Doctorates, 2006 was the fifth year in a row in which the majority of research PhDs awarded to U.S. citizens went to women.[7] Women earn more PhDs than men in the humanities, social sciences, education, and life sciences[8] and now serve as presidents of Harvard, MIT, Princeton, the University of Pennsylvania, and other leading research universities.[9]

Elsewhere, though, the figures are different. Women comprise just 28 percent of tenure-track professors in math, 18 percent in physics, 20 percent in computer science, and 14 percent in electrical engineering.[10] And the pipeline does not promise statistical parity in these fields any time soon: Women are now earning just 25 percent of the PhDs in the physical sciences—way up from the 4 percent of the 1960s, but still far behind the rate at which they are winning doctorates in other fields. "The change is glacial," says Debra Rolison, a physical chemist at the Naval Research Laboratory.[11]

Rolison, who describes herself as an "uppity woman," has a solution. A popular anti–gender bias lecturer, she gives talks with titles like, "Isn't a Millennium of Affirmative Action for White Men Sufficient?" She wants to apply Title IX to science education.[12] Although this celebrated gender-equity provision of the Education Amendments Act of 1972 has so far mainly been applied to college sports, it is not limited to them. The measure provides, "No person in the United States shall, on the basis of sex . . . be denied the benefits of . . . any education program or activity receiving federal financial assistance."[13]

While Title IX has been effective in promoting women's participation in sports, it has also caused serious damage, in part because it has led to the adoption of a quota system. Over the years, judges, U.S. Department of Education officials, and college administrators have interpreted Title IX to mean that women are entitled to "statistical proportionality." That is to say, if a college's student body is 60 percent female, then 60 percent of the athletes should be female—even if far fewer women than men are interested in playing sports at that college. But many athletic directors have been unable to attract the same proportion of women as men. To avoid government harassment, loss of funding, and lawsuits, they have simply eliminated men's teams.[14] While many factors affect the evolution of men's and women's college sports, Title IX has unquestionably led to men's participation being calibrated to the level of women's interest. That kind of calibration could devastate academic science.

Unfortunately, in her enthusiasm for Title IX, Rolison is not alone. On October 17, 2007, a subcommittee of the U.S. House Committee on Science and Technology convened to learn why women are "underrepresented" in academic professorships of science and engineering, and to consider what the federal government should do about it.[15]

As a rule, women tend to gravitate to fields such as education, English, psychology, biology, and art history, while men are much more numerous in physics, mathematics, computer science, and engineering. Why this is so is an interesting question—and the subject of a substantial empirical literature. The research on gender and vocation is complex, vibrant, and full of reasonable disagreements; there is no single, simple answer.

There were, however, no disagreements at the congressional hearing. All five expert witnesses, and all five congressmen, Democratic and Republican, were in complete accord.[16] They attributed the dearth of women in university science to a single cause: sexism. And there was no dispute about the solution. All agreed on the need for a revolutionary transformation of American science itself. "Ultimately," said Kathie Olsen, deputy director of the National Science Foundation, "our goal is to transform, institution by institution, the entire culture of science and engineering in America, and to be inclusive of all—for the good of all."[17]

Representative Brian Baird, the Democrat from Washington State who chairs the Subcommittee on Research and Science Education, looked at the witnesses and the crowd of more than a hundred highly appreciative activists from groups like the American Association of University Women and the National Women's Law Center and asked, "What kind of hammer should we use?"[18]

For the five male, gray-haired congressmen, the hearing was a happy occasion—an opportunity to be chivalrous and witty before an audience of concerned women, and to demonstrate their goodwill and eagerness to set things right.[19] It was also a historic occasion—more than the congressmen may have realized. During the past thirty years, the humanities have been politicized and transformed beyond recognition. The sciences, however, have been spared. There seems to have been a tacit agreement, especially at the large research universities, that while activists and deconstructionists would be left relatively free to experiment with fields like comparative literature, cultural anthropology, communications, and, of course, women's studies, the hard sciences—vital to our economy, health, and security, and to university funding from the federal government, corporations, and the wealthy entrepreneurs among their alumni—were to be left alone. Departments of physics, math, chemistry, engineering, and computer science have remained traditional, rigorous, competitive, and relatively meritocratic, and under the control of no-nonsense professors dedicated to objective standards. All that may be about to change. Following years of meticulous planning by the activists who gathered for the hearing, the era of academic détente is coming to an end.

The first witness was Donna Shalala, president of the University of Miami and secretary of health and human services in the Clinton administration. She had chaired the "Committee on Maximizing the Potential of Women in Academic Science and Engineering," organized by several leading scientific organizations, including the National Academy of Sciences (NAS), the Academy of Engineering, and the Institute of Medicine. In 2006 the committee released a report, *Beyond Bias and Barriers: Fulfilling the Potential of Women in Academic Science and Engineering*, that claimed to find "pervasive unexamined gender bias." It received lavish media attention and became the standard reference work for the "STEM" (science, technology, engineering, and math) gender-equity movement.[20]

At the hearing, Shalala warned that strong measures would be needed to improve the "hostile climate" women face in the academy. This "crisis," as she called it, "clearly calls for a transformation of academic institutions. . . . Our nation's future depends on it."[21] She and other speakers called for rigorous application of Title IX and sanctions against uncooperative institutions. Witness Freeman Hrabowski, president of the University of Maryland, Baltimore County, stressed the need to threaten obstinate faculties with loss of funding: "People listen to money. . . . Make the people listen to the money talk!"[22]

The idea of "title-nining" academic science was first proposed by Debra Rolison in 2000. She has promoted Title IX as an "implacable hammer" guaranteed to get the attention of recalcitrant faculty.[23] Prompted by Rolison and a growing chorus of activists, the U.S. Senate Subcommittee on Science, Technology, and Space held a 2002 hearing on "Title IX and Science."[24] Later, in 2005, the former subcommittee chairmen, Senators Ron Wyden (Democrat of Oregon) and George Allen (Republican of Virginia), held a joint press conference with feminist leaders. Wyden declared, "Title IX in math and science is the right way to start."[25] Allen seconded, "We cannot afford to cut out half our population—the female population."[26] The Title IX reviews have already begun.

At the October 2007 subcommittee meeting, Representative Vernon Ehlers, a Michigan Republican and self-described "recovering sexist,"[27] cheerfully suggested we declare science a sport and then regulate it the way we do college athletics.[28] He was joking, but it is important to recognize that science is not a sport. The purpose of college sports is to develop the skills and confidence of young athletes and to promote school spirit, while the goal of science is to advance knowledge. Success in fields like math, physics, computer science, and engineering is critical to our national security and well-being.

There is another essential difference between sports and science: In science, men and women play on the same teams. Very few women can compete on equal terms with men in lacrosse, wrestling, or basketball; by contrast, there are many brilliant women in the top ranks of every field of science and technology, and no one doubts their ability to compete on equal terms. Yet a centerpiece of STEM activism is the idea that science, as currently organized and practiced, is intrinsically hostile to women and presents a barrier to the realization of their unique intellectual potential. MIT biologist Nancy Hopkins, an effective leader of the science equity campaign (and a prominent accuser of Harvard president Lawrence Summers, when he committed the solecism of suggesting that men and women might have different propensities and aptitudes), points to the hidden sexism of the obsessive and competitive work ethic of institutions like MIT.

"It is a system," Hopkins says, "where winning is everything, and women find it repulsive."[29] This viewpoint explains the constant emphasis, by equity activists such as Shalala, Rolison, and Olsen, on the need to transform the "entire culture" of academic science and engineering. Indeed, the charter for the October 17 congressional hearing placed primary emphasis on academic culture:

> The list of cultural norms that appear to disadvantage women . . . includes the favoring of disciplinary over interdisciplinary research and publications, and the only token attention given to teaching and other service during the tenure review process. Thus it seems that it is not necessarily conscious bias against women but an ingrained idea of how the academic enterprise "should be" that presents the greatest challenge to women seeking academic S&E [science and engineering] careers.[30]

When the women-in-sports movement was getting underway in the early 1990s, no one suggested that its success would require transforming the "culture of soccer," or putting an end to the obsession with competing and winning. The notion that women's success in science depends on changing the rules of the game seems demeaning to women—but it gives the STEM-equity movement extraordinary scope, commensurate with the extraordinary power that federal science funding would put at its disposal.

Already, the National Science Foundation (NSF) is administering a multi-million-dollar gender-equity program called ADVANCE, which, as Olsen told the subcommittee, aims to transform the culture of American science to make it gender-fair.[31] Through ADVANCE, the NSF is attempting to make academic science departments more cooperative, democratic, and interdisciplinary, as well as less obsessive and stressful.[32] Furthermore, a few weeks before the hearing, a "Gender Bias Elimination Act" was introduced by one of the subcommittee members, Representative Eddie Bernice Johnson (Democrat of Texas), that would mandate not only stringent Title IX reviews but also bias-awareness workshops for academics seeking government funding.[33]

These proposed solutions assumed the existence of a problem where there might not be one. During her presidential campaign, Hillary Clinton noted that "women comprise 43 percent of the workforce but only 23 percent of scientists and engineers" and insisted that government take "diversity into account when awarding education and research grants."[34] But what is the basis for this and other attempts to balance the statistics? If numerical inferiority were sufficient grounds for charges of discrimination or cultural insensitivity, Congress would be holding hearings on the crisis of underrepresentation of men in higher education. After all, women earn most of the degrees—practically across the board. What about male proportionality in the humanities, social sciences, and biology? The physical sciences are the exception, not the rule.

So why are there so few women in the physical sciences and the higher echelons of academic math? In a recent survey of faculty attitudes on social issues, sociologists Neil Gross of Harvard and Solon Simmons of George Mason University asked 1,417 professors what they believed accounted for the relative scarcity of female professors in math, science, and engineering. Just 1 percent of respondents attributed the scarcity to women's lack of ability, 24 percent to sexist discrimination, and 74 percent to differences in what characteristically interests men and women.[35]

Many experts who study male/female differences provide strong support for that 74 percent majority. Readers can go to books like David Geary's *Male, Female: The Evolution of Human Sex Differences* (1998), Steven Pinker's *The Blank Slate: The Modern Denial of Human Nature* (2002), and Simon Baron-Cohen's *The Essential Difference: The Truth about the Male and Female Brain* (2003) for arguments suggesting that biology plays a distinctive—but not exclusive—role in career choices.

Baron-Cohen is one of the world's leading experts on autism, a disorder that affects far more males than females. Autistic persons tend to be socially disconnected and unaware of the emotional states of others. But they often exhibit obsessive fixation on objects and machines. Baron-Cohen suggests that autism may be the far end of the male norm—the "extreme male brain," all systemizing and no empathizing. He believes that men are, "on average," wired to be better systemizers and women to be better empathizers.[36] It's a daring claim—but he has data to back it up, presenting a wide range of correlations between the level of fetal testosterone and behaviors in both girls and boys from infancy into grade school.

Harvard psychologist Marc Hauser has what seems to be the appropriate attitude about the research on sex difference: respectful and intrigued, but also cautious. When asked about Baron-Cohen's work, Hauser said, "I am sympathetic . . . and find it odd that anyone would consider the work controversial."[37] Hauser referred to research that shows, for example, that if asked to make a drawing, little girls almost always create scenes with at least one person, while boys nearly always draw things—cars, rockets, or trucks. And he mentioned that among primates, including our closest relations, the chimpanzees, males are more technologically innovative, while females are more involved in details of family life. Still, Hauser warns that a lot of seemingly exciting and promising research on sex differences has not panned out, and he urges us to treat the biological theories with caution.

Nevertheless, it is hard not to be attracted to theories like Simon Baron-Cohen's when one looks at the way men and women are distributed in the workplace. After two major waves of feminism, women still predominate—sometimes overwhelmingly—in such empathy-centered fields as early-childhood education, social work, veterinary medicine, and psychology, while men are overrepresented in the "systemizing" vocations such as car repair, oil drilling, and electrical engineering.

Rachel Maines, a visiting scholar in science and technology studies at Cornell University, recently wrote an essay expressing amazement with women's progress in veterinary medicine as compared with engineering. Nationally, women now comprise fully 77 percent of students in veterinary schools, compared with 8 percent in the 1960s.[38] Maines writes, "To be sure, puppies are cuter than microchips, but most of what veterinarians do isn't about cute. Veterinary medicine . . . remains irreducibly bloody, messy, and often hazardous. . . . It certainly requires a rigorous scientific education that is at least as difficult and daunting as what engineering demands."[39]

Maines is surprised that women have managed so rapidly to take over this male-centered, science-based field without the benefit of bias workshops or federal equity initiatives. Cornell, she notes, just received a $3.3 million grant from the NSF to build a "critical mass" of women in all the STEM disciplines—ASAP. It is a first principle of the equity movement that role models and mentors are essential for helping women to move ahead in a field. But where, asks Maines, were the mentors and role models in veterinary medicine? She urges her colleagues to investigate the mystery of what happened.[40]

Theorists like Baron-Cohen may have solved that mystery. If Baron-Cohen is right, veterinary medicine would be a dream job for the scientifically gifted but empathy-driven female. This challenging and exciting field appeals to the feminine propensity to protect and nurture—and the desire to work with living things. An immense amount of literature documents male and female differences in choice of vocation. It also goes without saying that a lot of women will defy the stereotype of their sex and gladly enter systemizing fields, free of people, children, or animals—professions like mechanical engineering, metallurgy, or agronomy. But the number of men eager to enter these fields is markedly greater.

Let us go back to Math 55 for a moment. Baron-Cohen, along with many other scholars who write about cognitive sex differences, would not be surprised to learn that the students who show up in Math 55 are overwhelmingly male. The Harvard registrar's office reports that as of 2006, a total of seventeen women had completed the course since 1990.[41] Still, the equity activists could be right when they claim that the few women who defy the stereotype and take such a course have to overcome a "chilly environment."

I located two female survivors—Sherry Gong, then currently enrolled, and Kelley Harris, who had completed Math 55 with an A the previous year. "Did you encounter a hostile environment in that class?" I asked Harris. She laughed. "I loved my classmates!" When she once thought of dropping out, it was her male friends in the course who persuaded her to stay. Sherry Gong was taken aback when inquired whether she felt that women in math were unwelcome or marginalized. It was as if I had asked whether women had the vote. "It is 2007!" she reminded me. Sergei Bernstein, a young man enrolled with Gong, told me, "We would like to have more girls."[42]

The research emphasizing the importance of biological differences in determining women's and men's career choices is not decisive, but it is serious and credible. So the question arises: How have so many officials at the NSF and NAS and so many legislators been persuaded that we are facing a science crisis that Title IX enforcement and gender-bias workshops can resolve?

The answer involves a body of feminist research that purports to prove that women suffer from "hidden bias." This research, artfully presented with no critics or skeptics present, can be persuasive. A brief look at it helps explain the mindset of the critics and their supporters. But it is a highly ironic story, for the three recognized canons of the literature are, in key respects, travesties of scientific method, and they have been publicized and promoted in ways that have ignored elementary standards of transparency and objectivity. If they are auguries of how the STEM-equity activists intend to transform the culture of science, the implications are deeply disquieting. We begin with the famous, and mysterious, MIT study.

In 1994, sixteen senior faculty women, led by biologist Nancy Hopkins, complained to the administration about sex discrimination in their various departments. MIT's president, Charles Vest, and the dean of the School of Science, Robert Birgeneau, dutifully set up a committee to review the complaints. But rather than bring in outsiders, they put the protesters (joined by three male administrators) in charge of investigating their own grievances. Under Hopkins's leadership, the committee produced a 150-page study that found MIT guilty on all counts. Faculty women, according to the document, had lower salaries, less laboratory space, and fewer resources than faculty men. They felt "invisible" and "marginalized."[43] Vest and Birgeneau quickly responded with generous salary raises, improved lab space, and more equity committees. The women professed to be satisfied, and the case was closed. The report was deemed "confidential" and "sensitive," and to this day it has never been made public.[44]

What was released to the press, in March of 1999, was a brief summary of the report's findings, along with letters from Vest and Birgeneau admitting guilt. As the *Chronicle of Higher Education* reported, "MIT released a cursory report of the study it conducted, so it is difficult for outsiders to judge what the gap was between men and women."[45] The summary of the report nevertheless created a sensation in the media and in universities for two reasons: First, it appeared to be based on hard data, and second, it had the full endorsement of MIT's top administrators. The *New York Times* carried the story on the front page under the headline, "M.I.T. Admits Discrimination Against Female Professors."[46] Professor Hopkins was soon everywhere in the press and, on April 8, 1999, was invited to attend an Equal Pay Day event at the White House.[47] Referring to Hopkins and her team, President Bill Clinton said, "Together they looked at cold, hard facts about disparity in everything from lab space to annual salary."[48]

But cold, hard facts had little to do with it. After reviewing the available evidence and interviewing some insiders, University of Alaska psychologist Judith Kleinfeld concluded, "The MIT report presents no objective evidence whatsoever to support claims of gender discrimination in laboratory space, salary, research funds, and other resources."[49] Readers are told in the summary report that women faculty "proved to be underpaid."[50] But we also learn that the "salary data are confidential and were not provided to the committee."[51] So on what basis did they conclude there were salary disparities? Hopkins and the other authors explained, "Possible inequities in salary are flagged by the committee from the limited data available to it."[52] But "possible" soon became "actual," and by the time they reached President Clinton they had morphed into "cold, hard facts."

There were other oddities. The report claimed that the problems confronting women faculty were universal, but the summary conceded, "Junior women felt included and supported by their departments."[53] Instead of acknowledging that the problem might be generational and confined to a small group of senior women from three departments, Hopkins and the other authors of the report claimed that the junior women were naïve and simply did not know what was in store for them: "Each generation of young women began . . . by believing that gender discrimination was solved in the previous generation and would not touch them."[54]

Mathematics professor Daniel Kleitman, one of the three males on the Hopkins committee, told the *Chronicle* that he "never saw any evidence" of discrimination against women. He conceded that the senior women were unhappy, and he did not fault the administrators for trying to remedy the situation. But, as he explained, one can find unhappy professors in all universities. "I am not sure what the women were experiencing was unique to women," he said.[55]

I recently asked Kerry Emanuel, an MIT professor in earth, atmospheric, and planetary science, about the report. He told me that although it was "widely praised in public, it was privately deplored and disparaged in the hallways of MIT." His department was accused of bias, so he expected to see the evidence. "But it was never made available."[56]

When a reporter from the *Chronicle of Higher Education* asked Mary-Lou Pardue, an MIT biology professor who was among those who originally complained to the dean, about all the irregularities and the absence of data, she replied, "This wasn't meant to be a study for the rest of the world. It was meant to be a study for us. . . . We weren't trying to prove anything to the world."[57]

But the world thought otherwise. Vest and Birgeneau gave the impression that the report presented solid factual evidence of pervasive gender bias. When a *Wall Street Journal* editorial faulted the study, the two sent a letter claiming that the work of their committee had "successfully identified the root causes of a fundamental failure in American academia."[58] Feminist groups like the National Women's Law Center and the American Association of University Women were electrified and got ready for action—and action they got. As a direct result of the MIT report, the Ford Foundation, along with an anonymous donor, came forward with grants in excess of $1 million to fund more equity studies and to promote more initiatives to fight gender bias in academic science—and then the NSF followed suit with its ADVANCE institutional transformation campaign.[59]

The second key study in the literature purporting to reveal bias against women in the sciences appeared in May 1997 in the distinguished British journal *Nature*, in a provocative article entitled, "Nepotism and Sexism in Peer-Review."[60] The authors, Christine Wennerås and Agnes Wold, two

Swedish scientists from the University of Göteborg, claimed to have found strong gender bias hiding in the peer-review system of the Swedish Medical Research Council. After reviewing the relevant data, they concluded that to win a postgraduate science fellowship, a female applicant had to be at least twice as good as a male applicant.[61]

The Wennerås-Wold article caused a sensation both in Europe and the United States and is now a staple in the gender-equity literature. A 2007 article in *Scientific American* referred to it as the one and only "thorough study of the real-world peer-review process" and judged its findings "shocking."[62] When the NSF polled nineteen institutions that had received gender-equity ADVANCE grants, it asked which materials "had proved the most effective in their institutional transformation projects?" The Wennerås-Wold study made it to the NSF short list of four must-read "top research articles."[63] The Shalala/NAS *Beyond Bias* report described the piece as a "powerful" tool for educating provosts, department chairs, and search committees about bias.[64] The charter for the October 17 House subcommittee hearing gave particular prominence to the Swedish study.[65]

But what did the article actually show? Wennerås and Wold investigated the peer-reviewing practices of the Medical Research Council in 1994 after they had both been denied postgraduate fellowships. When they sought to review the data on which the decisions were based, the council refused to grant them access, insisting the information was confidential. But the two researchers went to court and won the right to see the data.

The Swedish study, unlike the MIT report, was actually published, and it presented data and described its methodology. But there are serious grounds for skepticism: Once again, it was a case of women investigating their own complaints; furthermore, what they concluded seemed a little improbable. According to their calculations, to score as well as a man, a woman had to have the equivalent of three extra papers published in world-class science journals such as *Science* or *Nature*, or twenty extra papers in leading specialty journals such as *Radiology or Neuroscience*.

I sent the Swedish study to two research psychologists, Jerre Levy (professor emerita of the University of Chicago) and James Steiger (professor and director of the Quantitative Methods and Evaluation Program in the Department of Psychology and Human Development at Vanderbilt University) for their review. They both immediately zeroed in on a troubling methodological anomaly: In the statistical analysis of their data, Wennerås and Wold had run separate regressions for only one productivity variable at a time. Since it is unlikely that any single variable adequately characterizes academic productivity, the obvious approach would have been to enter several of the productivity variables into a single regression equation. In any event, the dramatic results of the factor-by-factor approach that Wennerås and Wold used should have been tested against the more inclusive, realistic approach.

Steiger wrote to Wennerås and Wold requesting copies of the data so he could review them himself. Wold wrote back that she would gladly send the data, except that they had gone missing: "They were in a computer of a guy at the Statistics department and I got them on a diskette many years ago and I am afraid I will not be able to find it anymore."[66] Wennerås did not reply at all.

Certainly, researchers lose data. But these were pretty special data: The researchers had invested the substantial time and expense of a lawsuit to obtain them, and they were the basis of a highly celebrated study with singular findings.

But even assuming that the research held up, it is odd that a single study of postgraduate fellowships at a Swedish university should play such a prominent role in a campaign to eliminate "hidden bias" in American universities. Is it twice as hard for women as for men to receive postgraduate fellowships in the science departments of Berkeley or the University of Miami? If it is, would it not be straightforward to demonstrate the problem through at least one good study—one that follows customary statistical procedures and can stand up to peer review?

In fact, the NSF did do a review of its own grant-review process in 1997, and it found no evidence of bias against women. In 1996, for example, it approved grants from approximately 30 percent of female applicants and 29 percent of male applicants.[67] A formal outside study—done in 2005 by the RAND Corporation and titled "Is There Gender Bias in Federal Grants Programs?"—reached the same conclusion: "Overall, we did not find gender differences in federal grant funding outcomes in this study."[68] But, unlike the Swedish study, the RAND study did not make it to the NSF/NAS list of essential literature on gender bias. Two other items that were included in the "top four" are weak statistical studies of marginal issues that have never been rigorously evaluated.

A final item in the STEM-equity canon—the third of the three key studies mentioned above—is a book by feminist Virginia Valian that purports to be scientific, but is not. Virginia Valian, a psychologist at Hunter College, is one of the most-cited authorities in the crusade to achieve equity for women in the sciences. Her 1998 book *Why So Slow?* became indispensable to the movement because it offered a solution to a vexing problem: women's seemingly free but actually self-defeating choices. Not only do fewer women than men choose to enter the physical sciences, but even those who do often give child care and family a higher priority than their male colleagues. How, in the face of women's clear tendencies to choose other careers and more balanced lifestyles, can one reasonably attribute the scarcity of women in science and engineering to unconscious bias and sexist discrimination? Valian showed the way.

Her central claim was that our male-dominated society constructs and enforces "gender schemas." A gender schema is an accepted system of beliefs about the ways men and women differ—a system that determines what suits each gender. Wrote Valian:

> In white, Western middle-class society, the gender schema for men includes being capable of independent, autonomous action . . . [and being] assertive, instrumental, and task-oriented. Men act. The gender schema for women is different; it includes being nurturant, expressive, communal, and concerned about others.[69]

Valian did not deny that gender schemas have a foundation in biology, but she insisted that culture can intensify or diminish their power and effect. Our society, she said, pressures women to indulge their nurturing propensities while it encourages men to develop "a strong commitment to earning and prestige, great dedication to the job, and an intense desire for achievement."[70] All this inevitably results in a permanently unfair advantage for men.

To achieve a gender-fair society, Valian advocated a concerted attack on conventional gender schemas. This included altering the way we raise our children. Consider the custom of encouraging girls to play with dolls. Such early socialization, she said, creates an association between being female and being nurturing. Valian concluded, "Egalitarian parents can bring up their children so that both boys and girls play with dolls and trucks. . . . From the standpoint of equality, nothing is more important."[71]

But what if our daughters are not especially interested in trucks, as almost any parent can attest (including me: when my son recently gave his daughter a toy train to play with, she placed it in a baby carriage and covered it with a blanket so it could get some sleep)? Not a problem, said Valian: "We don't accept biology as destiny. . . . We vaccinate, we inoculate, we medicate . . . . I propose we adopt the same attitude toward biological sex differences."[72] In other words, the ubiquitous female propensity to nurture should be treated as a kind of disorder.

Valian was intent on radically transforming society to achieve her egalitarian ideals. She also wanted to alter the behavior of successful scientists. Their obsessive work habits, single-minded dedication, and "intense desire for achievement" not only marginalized women but also might compromise good science. In 2004 she wrote, "If we continue to emphasize and reward always being on the job, we will never find out whether leading a balanced life leads to equally good or better scientific work."[73] A world where women (and resocialized men) earn Nobel Prizes on flextime seems implausible. Unfortunately, her highly speculative ideas about changing men's and women's gender schemas are not confined to feminist theory.

Why So Slow? is trumpeted on the NSF/NAS "top research" list, and Valian herself has inspired the NSF's ADVANCE gender-equity program.[74] In 2001, the NSF awarded Valian and her Hunter colleagues $3.9 million to develop equity programs and workshops for the "scientific community at large."[75] Should Congress pass the Gender Bias Elimination Act, which mandates workshops for university department chairs, members of review panels, and agency program officers seeking federal funding, Valian will become one of the most prominent women in American scientific education.

The NSF has an annual budget of $6 billion devoted to promoting "the progress of science" and securing "the national defense."[76] It is not easy to understand how its ADVANCE program or its deep association with Virginia Valian is serving those goals.

In a 2005 interview, Alice Hogan, former director of ADVANCE, explained that the MIT study had been a wake-up call for the NSF. In the past, she said, the NSF had funded programs to support the careers of individual women scientists, but the MIT report persuaded its staff that "systemic" change was imperative.[77] Since 2001, the NSF has given approximately $107 million to twenty-eight institutions of higher learning to develop transformation projects. Hunter College, the site of Valian's $3.9 million program, is one of them. The University of Michigan has received $3.9 million; the University of Puerto Rico at Humacao, $3.1 million; the University of Rhode Island, $3.5 million; and Cornell, $3.3 million.[78] What are these schools doing with the money?

Some of the funds are being used for relatively innocuous, possibly even beneficial, projects such as mentoring programs and conferences. But there are worrisome programs, as well. Michigan is experimenting with "interactive" theater as a means of raising faculty consciousness about gender bias. At special workshops, physicists and engineers watch skits where overbearing men ride roughshod over hapless but obviously intellectually superior female colleagues.[79] The director/writer, Jeffrey Steiger of the University of Michigan theater program, explains that the project was inspired by Brazilian director Augusto Boal's 1979 book, *Theatre of the Oppressed*. Boal wrote, "I believe that all the truly revolutionary theatrical groups should transfer to the people the means of production in the theater."[80] To this end, the Michigan faculty members don't just watch the plays, but are encouraged to interact with the cast and even join them onstage. Some audience members will find the experience "threatening and overwhelming," so Steiger aims to provide them a "safe" context for expressing themselves.[81]

The NSF showcases this program as a "tried and true" success story. Michigan is not alone in using theater to advance the progress of science. The University of Puerto Rico at Humacao devoted some of its NSF–ADVANCE grant to cosponsor performances of Eve Ensler's raunchy play, *The Vagina Monologues*, a celebration of women's intimate anatomy.[82] The University of Rhode Island lists among its ADVANCE "events" a production of *The Vagina Monologues*, along with a visit by Virginia Valian. Rhode Island change agents, led by psychologist Barb Silver, are also trying to affect institutional transformation with a program called TTM—"Transtheoretical Model for Change." The program, adapted from one used by clinicians to help patients overcome bad habits and addictions such as smoking, overeating, and taking drugs, aims to break the Rhode Island faculty of its addiction to "traditional gender assumptions" and sexist behavior.[83]

Other schools are using their ADVANCE funds more conventionally— to initiate quota programs. At Cornell, as of 2006, twenty-seven of fifty-one science and engineering departments had fewer than 20 percent women faculty, and some had no women at all. The university is using its NSF grant for a program called ACCEL (Advancing Cornell's Commitment to Excellence and Leadership), dedicated to filling science faculty with "more than" 30 percent women in time for the university's sesquicentennial in 2015.[84] Sensible people—emphatically including the no-nonsense types who become scientists and engineers—will be inclined to dismiss the ADVANCE programs, the enthusiastic promotion of dubious research, and the well-intentioned but overly eager senators and congressmen, as an inconsequential sideshow in the onward march of mighty American science and technology. The NSF, like any government agency with a budget of $6 billion, can be expected to spill a few million here and there on silly projects and on appeasing noisy constituent groups. Unfortunately, the STEM-equity campaign is not going to rest with a few scientific bridges-to-nowhere.

For one thing, the Title IX compliance reviews are already underway. In the spring of 2007, the U.S. Department of Education evaluated the Columbia University Physics Department. Cosmology professor Amber Miller, talking to *Science* magazine, described the process as a "waste of time." She was required to make an inventory of all the equipment in the lab and indicate whether women were allowed to use various items. "I wanted to say, leave me alone, and let me get my work done."[85] But Miller and her fellow scientists are not going to be left alone. Most academic institutions are dependent on federal funding, and scientists like Miller and her colleagues can be easily hammered.

Equally worrisome is the fact that the NSF and NAS—America's most prestigious and influential institutions of science—have already made significant concessions to the STEM-equity ideology. So have MIT and Harvard. Can Cal-Tech be far behind?

The power and glory of science and engineering are that they are, adamantly, evidence-based. But the evidence of gender bias in math and science is weak at best, and the evidence that women are relatively disinclined to pursue these fields at the highest levels is serious. When the bastions of science pay respectful attention to the weak and turn a blind eye to the serious, it is hard to maintain the view that the scientific enterprise is somehow immune to the enthusiasms that have harmed other, supposedly "softer," academic fields.

Few academic scientists know anything about the equity crusade. Most have no idea of its power, its scope, and the threats they may soon be facing. The business community and citizens at large are completely in the dark. This is a quiet revolution. Its weapons are government reports that are rarely seen; amendments to federal bills that almost no one reads; small, unnoticed, but dramatically consequential changes in the regulations regarding government grants; and congressional hearings attended mostly by true believers.

American scientific excellence is a precious national resource. It is the foundation of our economy and of the nation's health and safety. Norman Augustine, retired CEO of Lockheed Martin, and Burton Richter, Nobel Laureate in Physics, once pointed out that MIT alone—its faculty, alumni, and staff—started more than five thousand companies in the past fifty years.[86] Will an academic science that is quota-driven, gender-balanced, cooperative rather than competitive and less time-consuming produce anything like these results? So far, no one in Congress has even thought to ask.

Notes

1. Mathematics Course Listing, Department of Mathematics, Harvard University: http://www.math.harvard.edu/pamphlets/freshmenguide.html.

2. Ury 2006.

3. Mathematics Course Listing, Department of Mathematics, Harvard University: http://www.math.harvard.edu/pamphlets/freshmenguide.html.

4. Ury 2006.

5. Ibid.

6. U.S. Department of Education, Institute of Education Sciences, National Center for Education Statistics, n.d.

7. Steward 2006.

8. Ibid.

9. The president of Harvard University is Drew Gilpin Faust. Susan Hockfield is president of MIT, and Shirley M. Tilghman is president of Princeton University. Amy Gutman serves as president of the University of Pennsylvania. For others, see American Council on Education 2004.

10. Nelson 2007, 14.

11. Rolison 2006.

12. Ibid.

13. "Section 1681 (a): Prohibition against discrimination; exceptions. No person in the United States shall, on the basis of sex, be excluded from participation in, be denied the benefits of, or be subjected to discrimination under any education program or activity receiving Federal financial assistance." U.S. Department of Labor, Office of the Assistant Secretary for Administration and Management 1972.

14. Epstein 2005.

15. U.S. House of Representatives, Committee on Science and Technology, Subcommittee on Research and Science 2007b.

16. Ibid. The witnesses included Dr. Donna Shalala, president, University of Miami; Dr. Kathie Olsen, deputy director, National Science Foundation; Dr. Freeman Hrabowski, president, University of Maryland; Dr. Myron Campbell, chair of physics, University of Michigan; and Dr. Gretchen Ritter, professor of government, University of Texas at Austin.

17. Olsen 2007.

18. Author's notes from hearing. See also Redden 2007.

19. Author's notes from hearing.

20. National Academy of Sciences et al. 2007, 4–31.

21. Author's notes from hearing. See also Consortium of Social Science Associations (COSSA) 2007.

22. Author's notes from hearing.

23. Rolison 2004.

24. Ibid. The hearing was held by the U.S. Senate on October 3, 2002.
25. Wyden 2005.
26. Ibid.
27. Rolison 2004.
28. Author's notes on hearing. See also Kasic 2007.
29. Kofol 2003.
30. U.S. House of Representatives, Committee on Science and Technology, Subcommittee on Research and Science 2007a, 5.
31. Olsen 2007.
32. National Science Foundation 2008b.
33. U.S. House of Representatives 2007.
34. HillaryClinton.com 2007.
35. Gross and Simmons 2007.
36. Baron-Cohen 1999. See also chapter 1, above.
37. Baron-Cohen 2005.
38. Maines 2007.
39. Ibid.
40. Ibid.
41. Ury 2006.
42. Interview by Christina Hoff Sommers, November 6, 2007.
43. *MIT Faculty Newsletter* 1999.
44. Ibid.
45. Wilson 1999.
46. Goldberg 1999.
47. Massachusetts Institute of Technology 1999.
48. Halber 1999. See also *MIT Tech Talk* 43, no. 26.
49. Kleinfeld 1999.
50. Ibid.
51. Committee on Women Faculty 1999.
52. Ibid., 8.
53. Ibid., 8.
54. Ibid., 9.
55. Wilson 1999.
56. Interview by Christina Hoff Sommers, November 7, 2007.
57. Smallwood 2001.
58. Vest and Birgeneau 2000.
59. Wilson 1999; National Science Foundation 2008b.
60. Wennerås and Wold 1997.
61. Ibid.
62. Halpern et al. 2007.
63. National Academy of Sciences et al. 2007, section 4, 37.
64. Ibid.

65. U.S. House of Representatives, Committee on Science and Technology, Subcommittee on Research and Science. 2007a, 5.

66. Email from Agnes Wold to James Steiger, November 16, 2007. For further discussion of the Swedish study, see chapter 3, above, and chapter 9, below.

67. Altman 1997.

68. RAND Corporation 2005.

69. Valian 1998, 13.

70. Ibid., 268.

71. Ibid., 30.

72. Ibid., 67.

73. Valian 2004, 8.

74. Stewart et al. 2007, 4.

75. Gender Equity Project n.d.

76. National Science Foundation 2008a.

77. Stewart et al. 2007.

78. National Science Foundation 2008b.

79. National Academy of Sciences et al. 2007, 4–24.

80. Boal 1979, 122.

81. Stewart et al. 2007, 221.

82. ADVANCE Institutional Transformation 2006.

83. ADVANCE University of Rhode Island 2007. See also Silver et al. 2006.

84. Steele et al. 2006.

85. Bhattacharjee 2007, 1776.

86. See MIT Entrepreneurship Center n.d.

References

ADVANCE Institutional Transformation. 2006. Activities. University of Puerto Rico at Humacao. http://www.uprh.edu/~advance/Recent_Activities.html (accessed January 13, 2009).

ADVANCE University of Rhode Island. 2007. Past Events Year Two. http://www.uri.edu/advance/news_and_calendar/past_events_year_two.html (accessed January 13, 2009).

Altman, Lawrence. 1997. Swedish Study Finds Sex Bias in Getting Science Jobs. *New York Times*. May 22.

American Council on Education. 2004. Women College Presidents Call for Greater Investment in Higher Education. August 26. http://www.acenet.edu/AM/Template.cfm?Section=Home&TEMPLATE=/CM/ContentDisplay.cfm&CONTENTID=13903 (accessed January 5, 2009).

Baron-Cohen, Simon. 1999. The Extreme-Male-Brain Theory of Autism. In *Neurodevelopmental Disorders*, ed. Helen Tager-Flusberg. Cambridge, Mass.: MIT Press.

———. 2003. *The Essential Difference: The Truth about the Male and Female Brain*. New York: Perseus Books Group.

———. 2005. The Assortative Mating Theory: A Talk with Simon Baron-Cohen. Edge: The Third Culture. April 4. http://www.edge.org/3rd_culture/baron-cohen05/baroncohen05_index.html (accessed January 5, 2009).

Bhattacharjee, Yudhijit. 2007. U.S. Agencies Quiz Universities on the Status of Women in Science. *Science* 315 (5820): 1776.

Boal, Augusto. 1979. *Theatre of the Oppressed*. New York: Theatre Communications Group.

Committee on Women Faculty. 1999. A Study of the Status of Women Faculty in Science at MIT. Massachusetts Institute of Technology. http://web.mit.edu/fnl/women/women.pdf (accessed January 6, 2009).

Consortium of Social Science Associations (COSSA). 2007. House Science Panel Examines Barriers to Recruitment and Retention of Women Faculty in Science and Engineering Fields. *COSSA Washington Update* 26 (19), October 22. http://www.cossa.org/volume26/26.19.pdf (accessed January 5, 2009).

Epstein, David. 2005. Title IX Guidance Stirs Debate. *Inside Higher Ed News*. March 23. http://www.insidehighered.com/news/2005/03/23/titleix (accessed January 13, 2009).

Geary, David C. 1998. *Male, Female: The Evolution of Human Sex Differences*. Washington, D.C.: American Psychological Association.

Gender Equity Project. n.d. The Gender Equity Project. Brochure. Hunter College. http://www.hunter.cuny.edu/genderequity/equityMaterials/GEPBascicBrochureforWeb.pdf (accessed January 13, 2009).

Goldberg, Carey. 1999. M.I.T. Admits Discrimination against Female Professors. *New York Times*. March 23. http://query.nytimes.com/gst/fullpage.html?res=9801E5DA1131F930A15750C0A96F958260 (accessed January 5, 2009).

Gross, Neil, and Solon Simmons. 2007. The Social and Political Views of American Professors. Working paper. September 24. http://www.wjh.harvard.edu/~ngross/lounsbery_9-25.pdf (accessed January 5, 2009).

Halber, Deborah. 1999. Clintons Praise MIT for Candor, Actions on Women Science Faculty. MIT news office. April 14.

Halpern, Diane F., Camilla P. Benbow, David C. Geary, Ruben C. Gur, Janet Shibley Hyde, and Morton Ann Gernsbacher. 2007. Sex, Math, and Scientific Achievement. *Scientific American.* November 28. http://www-rcf.usc.edu/~angelakn/Halpern07 SciAmMindGenderGapSci.pdf (January 5, 2009).

HillaryClinton.com. 2007. Hillary Clinton's Innovation Agenda: Rebuilding the Road to the Middle Class. Press release. October 10. http://www.hillaryclinton.com/news/release/view/?id=3656 (accessed January 5, 2009).

Hurder, Stephanie. 2008. Harvard Mathematics Department 21, 23, 25, or 55? Course pamphlet. Department of Mathematics. Harvard University. http://www.math.harvard.edu/pamphlets/freshmenguide.html (accessed January 5, 2009.

Kasic, Allison. 2007. The Coming Academic Title Wave. Independent Women's Forum. October 31. http://www.iwf.org/campus/show/19825.html (accessed January 6, 2009).

Kleinfeld, Judith S. 1999. MIT Tarnishes Its Reputation with Gender Junk Science. December 14. http://www.uaf.edu/northern/mitstudy (accessed January 6, 2009).

Kofol, Anne K. 2003. See No Evil. *Harvard Crimson,* June 5. http://www.thecrimson.com/printerfriendly.aspx?ref=348212 (January 6, 2009).

Maines, Rachel. 2007. Why Are Women Crowding into Schools of Veterinary Medicine But Are Not Lining Up to Become Engineers? *Cornell Perspectives.* June 12. http://www.news.cornell.edu/stories/June07/women.vets.vs.eng.sl.html (accessed January 6, 2009).

Massachusetts Institute of Technology. 1999. MIT Reports to the President 1998–1999. http://web.mit.edu/annualreports/pres99/17.00.html (accessed January 13, 2009).

MIT Entrepreneurship Center. n.d. MIT Spin-Offs. http://entrepreneurship.mit.edu/mit_spinoffs.php (accessed January 13, 2009).

MIT Faculty Newsletter. 1999. A Study on the Status of Women Faculty in Science at MIT. *MIT Faculty Newsletter* 11 (4). http://web.mit.edu/fnl/women/women.html (accessed January 6, 2009).

Mit Tech Talk, 43, no. 26.

National Academy of Sciences, National Academy of Engineering, and Institute of Medicine. 2007. *Beyond Bias and Barriers: Fulfilling the Potential of Women in Academic Science and Engineering.* Washington, D.C.: National Academies Press.

National Science Foundation. 2008a. About the National Science Foundation: NSF at a Glance. July 10. http://www.nsf.gov/about (accessed January 13, 2009).

———. 2008b. ADVANCE: Increasing the Participation and Advancement of Women in Academic Science and Engineering Careers. Updated November 10. http://www.nsf.gov/funding/pgm_summ.jsp?pims_id=5383 (accessed January 13, 2009).

Nelson, Donna J. 2007. A National Analysis of Minorities in Science and Engineering Faculties at Research Universities. October 31, 14, http://cheminfo.ou.edu/~djn/diversity/Faculty_Tables_FY07/07Report.pdf (January 6, 2009).

Olsen, Kathie L. 2007. Testimony Before the Committee on Science and Technology Subcommittee on Research and Science Education. National Science Foundation. October 17. http://www.nsf.gov/about/congress/110/klo_academicscieng_101707 .jsp (accessed January 13, 2009).

Pinker, Steven. 2002. *The Blank Slate: The Modern Denial of Human Nature.* New York: Viking Penguin.

RAND Corporation. 2005. Is There Gender Bias in Federal Grant Programs? Research brief. http://www.rand.org/pubs/research_briefs/RB9147/index1.html (accessed January 13, 2009).

Redden, Elizabeth. 2007. Female Faculty and the Sciences. *Inside Higher Ed.* October 18. http://www.insidehighered.com/news/2007/10/18/womensci (accessed January 6, 2009).

Rolison, Debra R. 2004. Title IX as a Change Strategy for Women in Science and Engineering . . . and What Comes Next. Barnard Center for Research on Women. December 9–10. http://www.barnard.edu/bcrw/womenandwork/rolison.htm (accessed January 23, 2009).

———. 2006. The Change is Glacial. *Chronicle of Higher Education.* May 26. http://chronicle.com/free/v52/i38/38a01101.htm (accessed January 23, 2009).

Silver, Barbara, et al. 2006. A Warmer Climate for Women In Engineering at the University of Rhode Island. American Society for Engineering Education. 2006 Annual Conference Proceedings Paper. June. Available: http://www.uri.edu/advance/files/pdf/ASEE_Final_Draft.pdf.

Smallwood, Scott. 2001. Report Questions MIT's Study on Treatment of Female Professors. *Chronicle of Higher Education.* February 16. http://chronicle.com/free/v47/i23/23a01701.htm (accessed January 23, 2009).

Steele, Bill. 2006. Cornell Launches $3.3 Million NSF Program to Build "Critical Mass" of Women Faculty in Engineering and Sciences. *Chronicle Online.* Cornell University. September 25. http://www.news.cornell.edu/stories/Sept06/NSFtransformation.ws.html (accessed January 23, 2009).

Steward, Doug. 2006. Report on the Survey of Earned Doctorates, 2006. http://www.ade.org/reports/SED_2006.pdf (accessed January 23, 2009).

Stewart, Abigail J., Janet Malley, and Danielle LaVague, eds. 2007. *Transforming Science and Engineering: Advancing Academic Women.* Ann Arbor, Mich.: University of Michigan Press.

Ury, Logan R. 2006. Burden of Proof. *Harvard Crimson,* December 6. http://www.thecrimson.com/article.aspx?ref=516216 (accessed January 23, 2009).

U.S. Department of Education. Institute of Education Sciences. National Center for Education Statistics. n.d. Fast Facts. http://nces.ed.gov/fastfacts/display.asp?id=93 (accessed January 13, 2009).

U.S. Department of Labor. Office of the Assistant Secretary for Administration and Management. Title IX, Education Amendments of 1972. http://www.dol.gov/oasam/regs/statutes/titleIX.htm (accessed January 20, 2009).

U.S. House of Representatives. 2007. Gender Bias Elimination Act of 2007. HR 3514. 110th Cong., 1st sess. (September 10, 2007), http://www.opencongress.org/bill/110-h3514/show (accessed January 20, 2009).

———. Committee on Science and Technology. Subcommittee on Research and Science Education. 2007a. *Hearing Charter: Women in Academic Science and Engineering.* http://democrats.science.house.gov/Media/File/Commdocs/hearings/2007/research/17oct/hearing_charter.pdf (accessed January 13, 2009).

———. Committee on Science and Technology. Subcommittee on Research and Science Education. 2007b. Subcommittee Investigates Barriers to Women Seeking Science and Engineering Faculty Positions. Press release. October 17. http://science.house.gov/press/PRArticle.aspx?NewsID=2000 (accessed January 13, 2009).

Valian, Virginia. 1998. *Why So Slow? The Advancement of Women.* Cambridge, Mass.: MIT Press.

———. 2004. Beyond Gender Schemas: Improving the Advancement of Women in Academia. *NWSA Journal* 16 (1): 207–20.

Vest, Charles M., and Robert J. Birgeneau. 2000. Vest, Birgeneau Answer News Critique of MIT Gender Study. MIT News Office. January 12. http://web.mit.edu/newsoffice/2000/wsj-0112.html (accessed January 8, 2009).

Wennerås, Christine, and Agnes Wold. 1997. Nepotism and Sexism in Peer-Review. *Nature* 387: 341–43. http://www.nature.com/nature/journal/v387/n6631/abs/387341a0.html (accessed January 13, 2009).

Wilson, Robin. 1999. An MIT Professor's Suspicion of Bias Leads to a New Movement for Academic Women. *Chronicle of Higher Education.* December 3.

Wyden, Ron, Office of. 2005. Wyden, Allen Accept Call to Break Down Barriers for Women and Girls in Math, Hard Science Fields. Press release. May 11. http://wyden.senate.gov/newsroom/record.cfm?id=266285 (accessed January 13, 2009).

5

Low Numbers: Stereotypes and the Underrepresentation of Women in Math and Science

Joshua Aronson

Not long after Harvard president Lawrence Summers was publicly excoriated and ultimately removed from office for sharing his thoughts about men being naturally better scientists and mathematicians than women, I found myself invited to sit on various panels convened to "debate the scientific evidence" on gender differences in math and science, specifically to present my research on "stereotype threat," which is the psychological predicament women and other minority group members face when confronted by cultural stereotypes alleging lower intellectual ability. Nearly all of these symposia were alike in their titles, which were questions, like this:

Are Women Being Held Back in Math and Science?

Female Under-Representation in Science: Nature or Nurture?

Was Larry Summers Right about Sex Differences in Math and Science?

As provocative as such questions are, they have the unfortunate tendency of eliciting answers that are either wrong or ultimately of little use. Indeed,

The author is thankful to Claude Steele and Stacey Rosenkrantz Aronson for helpful comments, and to the National Science Foundation, the Spencer Foundation, the Russell Sage Foundation, and the William T. Grant Foundation for supporting the research reported in his chapter.

even the notion of a "debate" on the scientific issues presumes that the number of positions one can take on a given issue is two, and thus leads us to ignore the middle ground, the complex ways nature and nurture interact. Judging from the apparent number of changed minds at the end of these debates—approximately zero—it is reasonable to ask whether the way we debate science may actually slow progress toward narrowing group differences and increasing the ranks of women in the highest levels of math and science.

It doesn't help matters that such discussions nearly always are conducted in the wake of a public crucifixion of the Larry Summers variety. At this writing, the most recent notable to be brought down for racial insensitivity is the Nobel Laureate James Watson, who, in attributing Africa's problems to the lower innate intelligence of Africans, found his reputation, his book tour, and his job directing a research center all destroyed by a few poorly chosen words. One need not agree with the Summerses or Watsons of the world to experience anguish at their professional demise, to cringe at the bigotry on both sides of the nature/nurture divide, and to shudder at the thinness of the political ice that surrounds us all, just waiting for a slip of the tongue.

If these debates have convinced me of anything, it is that the most malign influence is neither "IQ fundamentalism"[1] on the one hand, nor "political correctness" on the other. It's the *overconfidence* that so often accompanies either of these extremes. Whether it is expressed by a women's rights activist's fleeing the room at the mention of "biological differences" or the sneering hostility of a *Wall Street Journal* editorial extolling them, it is the overconfidence and dig-in-the-heels-and-say-anything-to-win mindset that keeps us debating rather than learning from the science on gender and race differences.

Although neither sneering nor shrill, the recent National Academy of Sciences report, *Beyond Bias and Barriers: Fulfilling the Potential of Women in Academic Science and Engineering*, is nonetheless a bit too confident about the barriers holding women back.[2] Its central argument is that girls and women are impaired by both conscious and unconscious biases that stem from cultural notions about female inferiority in math and science, and by processes such as "stereotype threat," in which one's own performance and motivation are spoiled by worries about living down to the low expectations these cultural notions impose. Although it is clear that gender bias exists and that stereotype threat—which I will describe in detail in this chapter—can significantly impede intellectual performance and development, the confidence that they play a big role in holding women back is unwarranted. Despite the admirable intentions of the NAS report, its authors are simply too confident that biological differences are not involved and that bias is. If we are true to the scientific process, we need to be clear on the fact that we do not know to what extent bias, stereotype threat, or other social factors contribute to the low numbers of women in math and science. Nor can we be confident that these processes *are not* involved.

The problem is that, unlike years of education, annual yearly income, or even IQ, constructs like gender bias and stereotype threat defy easy quantification; they cannot be easily entered into multiple regression analyses to calculate the percentage of variance explained in the gender gap—not in any terribly convincing way, at least. By the same token, we cannot confidently say how much biological "femaleness" matters in the gender gap, either. Thus, we are left knowing that biology and socialization matter, but how much each factor matters remains unclear.

This is precisely the conclusion reached by the psychologists Stephen Ceci and Wendy Williams in their recent book, *Why Aren't More Women in Science?*[3] After considering the research and viewpoints of the nation's sex-difference experts, they admit being unable to offer any firm conclusions about the relative roles played by nature and nurture in the low numbers of women in science. They further point out how the vast mosaic of findings that comprises the body of research on gender differences provides the scientific community with its own Rorschach test, affording diametrically opposed conclusions to be drawn—often from the same data—presumably depending upon one's implicit theories about nature and nurture, or upon one's political ideology. If we are rational and honest, the only reasonable response to such a mixed bag of scientific evidence like this is to get less confident, less extreme, and to become more tentative, to recognize that there are no simple answers. But studies of partisans reveal that people are often far more rationalizing than they are rational; and so mixed evidence has the paradoxical effect of further polarizing attitudes, leaving partisans ever more confident that they were right all along, and that those on the other side of the argument must be terribly biased. This is yet another reason these debates can be a waste of time.

As an applied social developmental psychologist, my read of the group-differences inkblot is greatly influenced by my goals as a researcher—which are not to focus on the question of *how much* group differences in achievement are caused by biology or socialization but, rather, on how an understanding of human psychology can help improve learning, motivation, performance, and intellectual development. This perspective renders the how-much-nature-how-much-nurture question irrelevant. Neither denying biological differences nor emphasizing them, I ask instead the practical question of whether differences among groups can be narrowed, and if so, how. Stereotype threat, a theory that I've been fortunate to help develop with my mentor, Claude Steele, and others, has been a fruitful vehicle for addressing these questions.

Stereotype Threat

A psychologist friend of mine spent a day at the White House some time ago to participate in a conference on programs to help inner-city children. President George W. Bush kicked off the proceedings with some introductory remarks. To my friend's surprise, the president spoke for nearly fifteen minutes without notes or teleprompter—and was articulate, fluent, and engaging. He was also genuinely witty and, at moments, even brilliant. In other words, he was utterly different from the person we have come to know from television, whose verbal gaffes fill the pages of such books as *Bushisms* and *The Bush Dyslexicon*, and whose behavior and speech have earned him derogatory nicknames and inspired countless late-night talk show jokes riffing on his assumed stupidity. My friend's experience with a surprisingly intelligent Bush was not an anomaly. After viewing a 1994 video of Bush's gubernatorial debate with Texas governor Ann Richards, a similarly astonished James Fallows wrote,

> This Bush was eloquent. He spoke quickly and easily. He rattled off complicated sentences and brought them to the right grammatical conclusions. He mishandled a word or two ("million" when he clearly meant "billion"; "stole" when he meant "sold"), but fewer than most people would in an hour's debate. More

striking, he did not pause before forcing out big words, as he so often does now, or invent mangled new ones.[4]

Many commentators, both Bush friends and Bush foes, have said that, off camera, he is far smarter and more articulate than people realize. Bush's SAT score, it turns out, places him in the same IQ range as John F. Kennedy, and higher than that of Bill Bradley and John Kerry. Stories friends tell of Bush's college days portray him as neither intellectually curious nor engaged, but reveal him as capable of stunning feats of intelligence—for example, being able to remember the names of a roomful of fifty fraternity brothers after hearing their names announced only once, something none of the other pledges could come close to doing. None of this means, of course, that Bush is an intellectual powerhouse whenever he's not on the spot, but it does raise an important question: What accounts for the fact that people can be demonstrably smart under some circumstances and yet apparently much stupider in others? What renders a person unable to demonstrate knowledge he or she can be assumed to have—say, on a test or during a speech? And what, if anything, might this have to do with racial and gender gaps in test scores?

The stock in trade of social psychology is to demonstrate that social context can exert powerful effects on human behavior and psychological functioning, overwhelming what we consider to be essential, defining personality characteristics. Hundreds of experiments demonstrate that the same individuals can act very differently—more competitively, more aggressively, more kindly—than their personality profiles or social reputations indicate, depending often upon small details of the social situation, the relationships among the actors involved, and the interpretations the actors draw about the context and the people in it.[5] Thus, "honest" students in one context have been shown to lie, cheat, and steal in another; a girl who disobeys her parents and dominates her younger brother in the home is shy, retiring, and obedient in school; the brilliant academic, adored by his students for his wit and charm in the classroom or on television, is awkward and boring in the presence of his mother-in-law—and so on. We've all experienced this, both as spectators of others' variability and as actors noticing our own. The research on stereotype threat simply extends this logic to the domain of human intellectual performance, and suggests that the predicament that has created difficulties for George Bush—a reputation for being ignorant and inarticulate that became widespread after he botched an on-air foreign policy quiz in 1999—can also operate at a group level, creating difficulties for individuals belonging to groups traditionally considered intellectually inferior. Human intellectual behavior is more fragile and malleable than we tend to realize, and what makes it fragile is that it is transacted within a web of social forces. These forces are powerful yet mostly invisible, and thus, when we evaluate the performances of presidents, college students, or fourth graders, we typically do a bad job of factoring them into our judgments of how smart people are.

Claude Steele and I tested the hypothesis that black college students may experience considerable apprehension when they are called upon to demonstrate their intelligence. In such situations (taking a test, speaking up in class, discussing a difficult concept with a fellow student, and so on), they face a predicament brought on by their awareness of the stereotype of African-American intellectual inferiority and their presumption that others may use the stereotype to evaluate them. Depending on the details of the situation, they may feel especially at risk of living up to the stereotype—of conforming to the image of the stereotypical stupid black person—and thus being devalued in the eyes of those around them. We further reasoned that the black student may even experience stereotype threat when no one else is watching, because the stereotype stands ready in his subconscious to explain any experienced difficulty performing an intellectual task. Such feelings could arouse anxiety, which, on complicated intellectual tasks, is known to disrupt performance, divide attention, and otherwise create an extra layer of unpleasantness not experienced by individuals for whom the stereotype does not apply. If this is true, we reasoned, perhaps it could help explain some part of the test-score gap—the portion that consistently remains after accounting for such factors as parents' level of intelligence, education, wealth, and the other "usual suspects" to which people customarily attribute group differences.[6]

We first tested these ideas with a series of laboratory experiments, in which African-American and white college students took a difficult verbal reasoning test. Half of the test-takers were given the test under normal testing conditions, with the test presented as a measure of verbal ability. For the remaining test-takers, we altered the test-taking situation, taking pains to assure the students that our purpose in having them solve the verbal items was not to measure their intelligence but, rather, to examine the psychology of problem-solving. The test performance results were quite striking. African Americans led to believe we were not interested in measuring their intelligence performed significantly better than those in the control condition. The difference in instructions had no appreciable effect on the white test-takers' performance.

Because black students were scarce at the university, we were forced to work with small numbers, and thus, despite a very strong effect, we needed to use a statistical correction to move from marginal to conventional statistical significance. Thus, we employed a commonly used statistical correction for test-takers' prior verbal scores on the SAT which adjusts the performances in a way that allows us to treat the students, who varied in their SAT scores, as though they had equivalent scores, thus putting us in a better position to disentangle the effects of experimental conditions from differences in test-taking ability and preparation. This correction is standard practice in such situations. Nonetheless, it has raised questions that have had the unfortunate effect of distracting some people from the essential message of this work.

The message, again, is this: Intellectual performance is more fragile and malleable than customarily thought, and attending to the psychological context—what it feels like to be an African American taking an IQ test, a women taking a math test, or George W. Bush making a speech with the television cameras rolling—can help us understand and improve performance and learning among individuals confronted by negative stereotypes or personal reputations. No matter how one analyzes the data, the conclusion is the same: We boosted African Americans' performance by deemphasizing an analysis of their ability, thereby making the situation psychologically safer and more conducive to intelligent thought. As one can see in figure 5-1, correcting for SAT or not correcting for SAT does not change this conclusion.

In the wake of the publication of this study in 1995, a good number of similarly constructed experiments (some three hundred at last count) have been conducted with targets of other group stereotypes—women taking mathematics tests, Latinos taking verbal tests, students of low socioeconomic status taking verbal tests, elderly individuals performing short-term memory tasks, women taking tests of political knowledge, chess, and driving, and even African Americans taking tests of miniature golf![7] In addition

FIGURE 5-1

VERBAL TEST PERFORMANCE UNCORRECTED BY SAT-V VS.
VERBAL TEST PERFORMANCE CORRECTED BY SAT-V

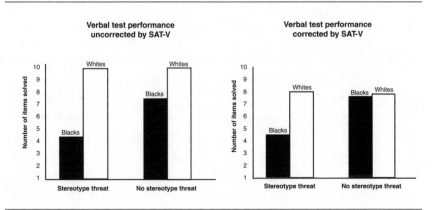

SOURCE: Test performance results from Steele and Aronson 1995.

to supporting the findings of the original study, some of these experiments also establish important insights about the nature of stereotype threat. Most important, perhaps, is the fact that one need not be regularly or persistently targeted by ability stereotypes to be affected by them. For example, my colleagues and I found that white males with very high math abilities can be led to choke on a test by informing them that their scores would be used to evaluate their ability relative to Asian test-takers.[8] Studies like this underscore the power of stereotypes to suppress intelligent thought and performance.

Reducing stereotype threat can improve performances that draw upon abilities thought to be mediated by hormones. For example, mental rotation—the ability to mentally rotate representations of two-dimensional or three-dimensional objects—has long been shown to produce robust gender gaps favoring males at all ages and has been linked in some studies to the amount of testosterone in the body.[9] This suggests a hard-wired sex difference in abilities and interests in spatial abilities with implications that range from a child's preference for playing with LEGOs over dolls, to performing well on geometry tests in high school, to success in engineering classes in college, all of which could help explain the low numbers of women in "hard" math and science careers. And yet Matthew McGlone and I found that we could significantly influence mental rotation performance on the Vandenberg Mental Rotation Test (a common measure of spatial ability) simply by influencing what test-takers thought about prior to the test. In our study, we gave male and female college students one of three mindsets. To get some of the test-takers thinking about their gender, we asked them to tell us a few things a student might like about living in a coed dorm. We compared this to a control condition in which they were asked to tell us a few things a student might like about living in the Northeast (where their college was located). As can be seen in figure 5-2, men performed better and women worse than controls when led to think about their gender. But we almost completely eliminated the gender difference in a third condition by getting them to think about a positive achieved identity—we asked them what someone might like about being a student at a "highly selective liberal arts college." Thus, psychological mindsets can overcome even robust sex differences in performance that are often attributed to biology.

In each case, the message of these studies is the same: Reduce stereotype threat and performance improves; induce it, and performance suffers. Many studies of this sort have also examined important individual differences in

FIGURE 5-2

IDENTITY SALIENCE INFLUENCES MENTAL ROTATION PERFORMANCE

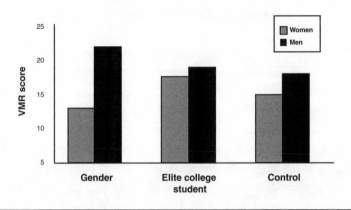

SOURCE: Vandenberg Mental Rotation Scores from McGlone and Aronson 2006.

what I call "stereotype vulnerability." Do students think that the stereotypes are true? Do they expect to face lots of discrimination? Do they care a great deal about academic success in a domain? Do they feel a strong connection to their racial or gender identity? Do they believe intelligence is essentially fixed and cannot develop? Students who answer yes to these questions tend to be more vulnerable to stereotype threat.

There is a fundamental point that must be stressed about such experiments—one that any student who has taken a course in experimentation understands, but is often lost on people not trained in social science. Laboratory experiments—either one or a group of three hundred—can tell us nothing about the degree to which a phenomenon occurs in the real world, or, indeed, if it occurs at all. This is not what experiments are designed to do. They simply tell us about average responses to conditions created in the experiment—how, say, a typical person processes information when he believes his intelligence is being measured. If people respond as we predict they will, we know that our theory of human information processing is correct, at least under the conditions we have specified and with the population we have sampled. If we are wrong, we revise our theories accordingly. The experiments on stereotype threat therefore cannot tell us *how much* of the race gap or the gender gap in achievement in the real world can be attributed to stereotype threat. The laboratory experiment is simply not designed to do this, any more than a hammer is designed to saw wood, turn screws, or apply paint.

This limitation, of course, applies to any experimental evidence under consideration in sex differences research. For example, Simon Baron-Cohen and his colleagues find that male newborns are particularly drawn to spatial objects—a mobile with parts that move mechanically—whereas female newborns are particularly interested in human faces.[10] These findings clearly establish, to my mind, a critical difference that stems from biology and is relevant to spatial abilities and, thus, math performance. But how much of the gender gap in math, or the low numbers of women who will later choose a career studying "things" (physics, computer science, or engineering) versus one oriented toward "people" (biology, sociology, or politics) can be attributed to these inborn differences? The data provide no information about how much of the gap is nature or how much later socialization will enter and change the picture—or any other sort of *how much* question about the gender gap in the real world. What they *do* tell us is this: Any theory attributing differences in interest in spatial tasks uniquely to socialization is either wrong or in need of revision, because these differences show up before any socialization can occur. That's the kind of work experiments do. They test theories about the way human minds work.

Stereotype threat experiments tell us, among other things, that contexts that arouse stereotype threat can have an impact on test scores. They also reveal the psychological processes involved, such as how the cognitive activation of stereotypes accompanies impaired performance and working memory capacity, as well as the activation of brain regions associated with social and physical threats.[11] They further suggest the conditions and mindsets that can reduce stereotype threat, such as being exposed to a role model—a female or black test administrator, for example—whose mere presence and assumed expertise can counter stereotype threat and raise the test scores of females and blacks taking the test.[12] Above all, at a most fundamental level, these experiments tell us that any theory holding that tests measure only innate intelligence or academic preparation are either wrong or in need of revision. It is the *understanding* gained from these experimental findings—namely, that intellectual performance is subject to social influences like stereotype threat—which we apply to the real world, not the exact experimental results themselves. Our understanding tells us that IQ and effort are not the only things that matter in test performance. Psychology matters, too. We think this is a useful thing to know.

Stereotype Threat in the Real World

To find out if and how stereotype threat plays a role in the real-world achievement of women and minority students, research must be conducted that tests these understandings of human performance in schools, in standardized testing centers, and in college classrooms. A number of studies have done just that. To find out how much of the gap in grade point average (GPA) may be attributed to stereotype threat, one approach is to measure stereotype vulnerability in students and predict their college performance over time. For example, Douglas Massey and Mary Fischer conducted a longitudinal survey of over four thousand freshmen from different ethnic backgrounds, attending more than twenty-eight American colleges.[13] Students were surveyed at the beginning of their college careers, and their performance was monitored thereafter. Large differences in GPA were found among ethnic groups; Asians and whites outperformed blacks and Latinos, even when controlling for SAT scores, family income, and other important background factors. The degree of measured stereotype vulnerability predicted 9–10 percent of their variation in grades, and, when accounted for in the predictive model along with the other background factors, explained the entire performance gap. In other words, stereotype threat explained the unexplained portion of the gap that was left over once we accounted for high school grades, family background, and prior measures of ability and preparation.

A second approach to examining the role of stereotype threat outside the laboratory involves asking whether women in the pipeline to math and science careers experience it. Students at the high end of the mathematics bell curve are frequently assumed to be immune to stereotype threat. To test this, Catherine Good, Jayne Anne Harder, and I examined the performances of women and men in the highest-level calculus classes offered at the University of Texas. All of the men and women in the class were informed by the professor that the test, like all of the tests in the course, was a measure of their calculus abilities and knowledge, but half were further informed that the test had never shown gender differences in the past, a framing of the test that had been shown to reduce stereotype threat in laboratory studies. In the control condition, where stereotype threat was not nullified, the women and men performed equally well, which is not surprising given that, on average, these women's grades were equal to the men's. But when stereotype threat was reduced by the statement about gender bias, the women performed significantly better than the men. Thus, stereotype threat appears to operate in high-level classes that future scientists must pass, apparently suppressing women's performances even on in-class tests.

A number of studies have gotten at the real-world applicability of the theory by conducting interventions based on the hypothesis that reducing feelings of stereotype threat in classrooms should increase grades and test scores. For example, Geoffrey Cohen and his colleagues, employing means that would cost a school system very little to implement, reduced the black–white grade gap among middle school students by 40 percent.[14] Cohen and colleagues did this by having students complete an in-class writing assignment that affirmed their most cherished values, an intervention shown in numerous laboratory studies to reduce or eliminate the effects of psychological threats by creating a sense of psychological safety. Catherine Good, Michael Inzlicht, and I conducted another theory-based intervention among middle school children and examined its effects on their statewide exit exam scores in reading and math.[15] The intervention involved having older students share their wisdom by teaching one group of younger students that human intelligence can grow with effort, and another group that most students improve their scores and grades once they get the hang of middle school. Both of these interventions, which had been shown to reduce test anxiety in the lab, significantly improved Latinos' reading scores and completely eliminated the gender gap in math test scores by improving the performance of the girls.

Considered together, the Cohen intervention and the Massey and Fisher longitudinal study of college students underscore a vital point—another that is missed by nonscientists: Even if stereotype threat accounts for, say, 9 percent of the achievement gap between two groups, interventions that reduce it can initiate a process that closes the gap to a much greater degree (for example, 40 percent). Reducing stereotype threat, in other words, may create the conditions that allow students to engage more deeply with their academic work and learn better, which in turn engages other processes, such as encouragement and praise from parents and teachers, more adaptive responses to critical feedback,[16] peer respect, and the love of learning—processes that have recursive, spiraling effects on achievement. For this reason, insisting, as some people have, on answering the how-much-of-the-gap question before doing anything about it strikes me as a misguided obsession. Knowing that factor x causes 10 percent of a problem does not force the conclusion that intervening upon that factor can produce an improvement of only 10 percent. In the case of stereotype threat and academic achievement, the size of the cause appears not to be isomorphic with the impact of intervening on that cause. It also means that what looks like a "quick fix" of reducing threat can open the door to "slower-fix processes," like greater willingness to study, an increased sense of valuing and enjoying learning, and better responses to feedback, all of which have positive effects of their own.

Such studies make me optimistic that gender and race gaps can be considerably narrowed, and that doing so requires neither massive nor expensive interventions. Small, timely, intelligently designed interventions can be as effective as, or even more effective than, the massive but ill-conceived interventions that are so often proposed. The failure of No Child Left Behind (NCLB) to reach its goals is a testament to the fact that interventions that misunderstand or ignore the psychology of human motivation are unlikely to work well, no matter how much legislative muscle they may have behind them. Charles Murray recently interpreted the failure of NCLB as evidence that essentially nothing can be done to reduce the achievement gap.[17] I read the failure very differently—not as proof of the intractability of the gap, but as a repudiation of the higher-test-scores-or-else approach to improving learning and motivation. Some years ago I predicted NCLB would fail for just this reason,[18] and data from recent large-scale experiments with the California Academic Skills Exit Exam bear this out.[19] The study found that requiring the exam for high school graduation lowered the minority students' performance across the board and girls' performance on the math test. Similarly skilled students performed better on the same exam a year earlier, when the stakes were lower. But just because you cannot force more learning and higher test scores with threats and high pressure does not mean you cannot get them through other, more thoughtful means, and our stereotype-threat-based interventions show this.

"Controversial" Issues in Stereotype Threat

Since I intend to play hardball in discussing the criticisms of stereotype threat, I should probably go up to bat first. There is no doubt about it: Steele and I made mistakes in our first publication on the test performance of African Americans. First, we overstated the situational side of the case, suggesting that the threat was located primarily "in the air," and not in some combination of person and social context. Our subsequent research quickly departed from this it's-all-in-the-situation stance. After all, reality is partly what people make of it, and therefore our individual personalities and mindsets—those that the culture imposes upon us, those we develop through life experiences, and those we choose—matter a great deal in how and how much we are affected by situations, such as the situations facing the targets of negative ability stereotypes. Still, many achievement situations are so powerful that a random sample of students is likely to do worse on a test when confronted with a stereotype—as in the study of male math whizzes confronted by the stereotype of Asian superiority, and the original Steele and Aronson study itself.[20] But I believe we were so excited about the power of the situation that we justifiably but insufficiently emphasized the theoretical role of individual differences in stereotype threat in that initial paper. A long list of studies illuminating these individual differences has corrected this error.[21]

A second mistake was that, in order to justify using prior SAT scores as a covariate, we made an odd argument, suggesting that stereotype threat probably did not affect students' SAT scores because the SAT was not sufficiently difficult for the Stanford University students included in the study. I no longer believe this. There is simply no theoretical reason why the SAT would be immune to stereotype threat if the test-taker experiences some difficulty with it. A recent meta-analysis of stereotype threat studies conducted in several countries and involving nearly 19,000 test-takers at all levels of ability suggests that on average, Latinos and blacks lose about 40 points on the verbal SAT due to stereotype threat, while women lose about 20 points on the math SAT.[22] In other words, because of stereotype threat, the SAT systematically underestimates the abilities of both women and non-Asian minorities.

Our third mistake, as noted earlier, is that we published our study with a graph of adjusted means that created the illusion that reducing stereotype threat eliminates the racial achievement gap, instead of presenting the unadjusted means as I have done in figure 5-1. Given all the misunderstandings this graph produced, I have come to think of that error—which appeared in only the third publication of my career, submitted some fifteen years ago—the same way I think of pictures of me in the 1970s, with my bad haircut, paisley shirts, and bell-bottom jeans: *What was I thinking?* At the same time, I'm comforted by the fact that science is a self-correcting enterprise. The errors and unanswered questions in that early paper have led other scientists to raise important questions and do their own research. And this is just as it should be: One study's flaws are addressed by another's strengths. One reason for the rapid growth of stereotype threat research is that our errors and loose ends have inspired so many independent laboratories to address the issues raised by our first publication on the subject and to do their own research. This is all for the good; it's how science progresses.

What has been most surprising about a few of the critics of stereotype threat research is that they have tended to focus almost exclusively on the very first study—the Steele and Aronson paper—in a series of three hundred, ignoring the fact that the theory has been greatly refined and reinforced by a corpus of subsequent research. I cannot think of another case in science, medicine, or technology where an entire body of informative research by independent laboratories around the world is called into question because of loose ends in the first study in a series. It is as if a few if-man-were-meant-to-fly fundamentalists in 1945 were harping about the future of aviation because of complications at Kitty Hawk.

The absurdity of this approach makes sense only when we consider where the criticisms are coming from—namely, nonscientists with an IQ fundamentalist agenda, or scientists connected with the testing industry.[23] Two implications of stereotype threat seem to bother them. First, if scores can be pushed around so easily by changes in the context, then cognitive tests may not be such great measures of intellectual worth. Second, they appear mightily resistant to the idea that institutions like schools can nurture intellectual development and make students smarter and more productive by attending to the psychology of achievement.

A conservative law professor named Amy Wax best exemplifies the IQ fundamentalist response to this research. Rather than conduct corrective research or sample the entire body of stereotype-threat research, she approaches scientific discussion with incuriosity and arguments more appropriate to a civil suit or an installment of *Hannity and Colmes* than to sober scientific discourse. In an op-ed piece she cites "grave methodological flaws" and the "lack of evidence" to support stereotype threat.[24] Don't get me wrong; she is right to raise questions about our research. The problem is that she is not interested in the answers—unless they confirm her prior beliefs, which she made clear in print before her public attacks on stereotype threat began:

> Some—perhaps many—people are not smart enough, or interested enough, to solve hard problems that demand the use of algebra and geometry. No amount of "educational reform" can change this.[25]

This is the IQ fundamentalist position on group differences: Nothing can be done to alter what is essentially set in genetic stone. The problem with this position (beyond the overconfidence and incuriosity) is simple: data.

Whether it's James Flynn's consistent finding that every succeeding generation's IQ has risen steadily in most cultures,[26] or the increasing ranks of women succeeding in mathematics and science,[27] or the fact that girls in Belgium and Japan score much higher in math and science than do boys in America,[28] or the fact that children randomly assigned to same-race,[29] same-sex,[30] or high-quality teachers[31] or to smaller classes[32] become measurably smarter and more engaged than their less fortunate peers, or the narrowing black–white test-score gap over several decades,[33] or any of our achievement-boosting interventions conducted under the aegis of stereotype threat— there is simply too much evidence that intellectual ability is shaped by culture and context for a rational social scientist to accept the notion that "no amount of school reform" can change the status quo. Indeed, as should be clear from the intervention research, even a small amount of school reform can do it; it just needs to be intelligently designed.

With IQ fundamentalism as one's working theory, one is forced in the face of contrary evidence either to revise the theory, or to twist or simply ignore the new evidence in order to preserve a sense of reasonableness. Wax has taken the latter approach in her treatment of stereotype threat.

For example, Wax bemoans the lack of evidence showing that reducing stereotype threat would improve the performance of women or minorities in the real world. Yet such evidence is abundant, as is noted above. At her request, I sent Wax articles showing that the grades and test scores of minority students and girls rise dramatically in response to simple interventions that reduce stereotype threat in schools and on statewide standardized exit tests, as well as in high-level college math courses. These were high-quality, randomized, controlled trials published in top journals—they cannot be written off either as laboratory demonstrations or as artifacts of statistical corrections. Wax made no mention of these promising effects in her critiques of stereotype threat, and, when confronted with the data during our October 2007 debate at the American Enterprise Institute (AEI), she simply changed the subject and moved on to another item in a litany of complaints. This is not how science progresses.

More recently, Charles Murray—a serious social scientist from whom I typically learn a lot and with whom I tend to agree on a great many issues—has dismissed the implications of this body of research as "educational romanticism." But to maintain such a position, he, too, must overlook data. He writes,

> The problem that gets in the way of this appealing story is that all of the experimental studies have explicitly induced a threat as part of the experiment's protocol. That threat consists of telling the experimental group that they are about to take a test that measures their innate ability. But tests in K–12 education are never presented that way.[34]

All of the experimental studies? In reality, studies exist (of which I made Wax and Murray aware) showing that stereotype threat is the default situation—where no mention is made about innate ability. In these experiments, test scores improve if test-takers are given instructions that reduce naturally occurring stereotype threat—for example, by presenting the test as a measure of an expandable ability, or as gender-fair, or as not measuring ability at all. A particularly striking example of this kind of study was conducted by the Educational Testing Service (ETS), which found it could reduce some of the stereotype threat in an actual AP calculus test simply by removing the customary request for gender information that precedes the test. This little change significantly boosted the performance of young women taking the calculus exam, and thereby significantly increased the number of women who qualified for AP credit. Indeed, according to one published analysis of the ETS data, simply moving the demographic information to the end of the actual test would "increase the number of women receiving AP calculus credit each year by 4,700."[35]

Wax and Murray are also unmoved by the longitudinal research on thousands of college students conducted by Massey and Fischer, showing that stereotype threat accounts for about 9 percent of the variance in college achievement among minority students.[36] This last point is particularly curious, because among Wax's strongest criticisms of the theory, voiced during her AEI presentation, is the claim that nobody has done any work showing how much of the achievement gap can be attributed to stereotype threat.[37] She said this only minutes after I presented such data.

Finally, and most depressingly, Wax has repeatedly misled the public regarding the nature and meaning of Paul Sackett's theoretical musings about the original Steele and Aronson research.[38] Sackett raises two very important points. First, he found that many media accounts and textbooks and a few journal articles had spoken of our research in a shorthand that implied the entire race gap was caused by and therefore could be eliminated by reducing stereotype threat. This misunderstanding was surely abetted by our graph showing only the SAT-corrected test scores without presenting the uncorrected scores. If true, this misperception would be unfortunate and indeed regrettable. Yet the rational, scientific approach to a myth is to look really closely at the data—all of them—and not simply generate a countermyth about our research being flawed.

Moreover, despite the alarm raised by Sackett's critique, I have never met, in years of contact with educators and psychology students, anyone who attributes the entire gap to stereotype threat. Professor Wax is the only person I've ever met who thinks that I have claimed this central role for the phenomenon, and repeatedly puts such claims in the mouths of stereotype threat researchers.[39] So, I'll be extra clear: *Stereotype threat does not explain the entire race gap*. To think so would require ignoring years of research on important contributing factors.[40] Indeed—and this is why the critiques are so surprising—the existence of stereotype threat does not even imply that there is no genetic or biological basis to group differences in ability; it simply means that among the many factors that contribute to group differences, it is an additional factor, one that appears to be significant, and highly amenable to intervention.

Sackett's second point is that by using an analysis of test scores that controlled for students' prior SAT scores, we possibly obscured the nature of the phenomenon. Specifically, he argues, we might have been showing that by jacking up stereotype threat in the laboratory, we could scare black students into scoring at lower levels than would be reflective of their "true" abilities (that is, as indicated by their SAT scores). But, he reasons, we were not showing that reducing stereotype threat, as we argued, could improve scores relative to the prior SAT scores. I should note that of the hundreds of experiments that confirm stereotype threat, only a handful use this correction.[41]

Nonetheless, Sackett's argument about the Steele and Aronson experiments is a reasonable one. But as compelling as it is, it advances a theoretical point, not an empirical one. In other words, Sackett is speculating about our data, not analyzing it. Our actual data did not support this reasoning; again, as shown in figure 5-1, when analyzed with or without the SAT correction, it is clear that black test-takers' scores were improved by assurances that we were not interested in measuring their intelligence. Still, to be certain of this requires a larger experiment than ours, one that presents the test in three ways rather than just two, so that we can determine if presenting the test as a measure of intelligence actually differs, as Sackett claims, from simply presenting it as one normally does—as a test, without adding the notion that we are examining intelligence and thus depressing scores by scaring students.

Ryan Brown and Eric Day conducted just such an experiment with students at the University of Oklahoma, and it shows that our interpretation was the correct one.[42] Brown and Day gave black and white college students the Raven's Progressive Matrices test, a culture-free, nonverbal IQ test that is prized by IQ researchers for being unbiased. One group of students was given the test with the official Raven's instructions (which do not mention intelligence). A second group (the stereotype-threat group) was told that the test was a measure of their intelligence, and a third (the stereotype-threat-reduced group) was told that the task was simply a puzzle. The results showed identical effects in the first two conditions: Blacks scored significantly worse than whites when they thought that the test was a measure of their intelligence or when the test was given under the standard Raven's instructions. Contrary to Sackett's theory, calling the Raven's a test of intelligence did not produce lower performances than did the standard instructions. As in the Steele and Aronson studies, the framing of the test in these two conditions did not affect the performance of the white students. Moreover, the patterns of data were identical whether or not the scores were adjusted by students' prior test scores. Thus, the argument that our experiment shows only that one can make blacks or women perform worse than their prior test scores by adding an artificial, surplus fear—one that doesn't exist in the real world of standardized testing—was refuted. In the "puzzle" condition, where stereotype threat was minimized, blacks performed dramatically better. These results perfectly confirm the notion that stereotype threat is operative when people take tests under standard testing conditions, and that making the situation less psychologically threatening can improve their test scores relative to their prior tested ability. The astute reader will note that this is just the sort of experiment Charles Murray claims does not exist.

Wax has repeatedly used the Sackett argument to impugn the stereotype threat research, despite being aware of research that demolishes its underlying premise. Worse, in her *Wall Street Journal* editorial, she pretends that Sackett and colleagues actually formulated their critique of our research with empirical data rather than speculation:

> As noted by University of Minnesota psychologist Paul Sackett and his colleagues in the January issue of *American Psychologist*, the raw, unadjusted scores of African-American and white students in the Steele/Aronson paper actually "differed to about the degree that would be expected on the basis of differences in prior SAT scores."[43]

The key words here are "the raw, unadjusted scores" and "actually," which she inserts to build a case that stereotype threat is simply a hothouse phenomenon, cooked up in the laboratory but with no meaning for the real world. The problem is that Sackett never actually examined the Steele and Aronson data. Had he requested them and analyzed them for himself, his argument would have collapsed. Raw or adjusted, the data show that reducing stereotype threat lifted African Americans' verbal test performance—just as they did with Brown and Day's Raven's scores, and just as such threat-reducing maneuvers have done in hundreds of studies with women and mathematics, both inside and outside the laboratory.

Needless to say, I find this kind of fudging deeply depressing—especially so when it is done in the name of arguing for innate and immutable group differences. Far from being a thoughtful, truth-seeking consideration of a diverse body of hundreds of published experiments, longitudinal field studies, and interventions, these critiques amount to little more than a scattershot venting of the spleen at the notion that there may be more to the story than the intellectual inferiority of blacks and the mathematical inferiority of women. I will leave it to experts on rationalization to explain why anyone would find so inconvenient the good news that simple and inexpensive means can be used to lift students to higher levels of engagement, enjoyment of academics, and intelligence.[44] But this attitude does underscore a most depressing feature of our debates of the evidence: Some—perhaps many—people are simply not capable of modifying deeply held beliefs about group differences—and no amount of data can change this.

Conclusion: The Grand Experiment

Are women being held back in math and science? Probably, but the extent to which they are held back by nurture versus nature remains unclear. The good news is that things are changing. True, women currently earn fewer than 25 percent of the bachelor's degrees in physics, engineering, and computer science—and only about a third of the PhDs in these areas.[45] But it is also true that this gap shows signs of closing. And for the past three decades, women have been catching up to men, even overtaking them, in many other areas of academics. In my own field of psychology, for example, women have gone from earning a tiny minority of degrees in the 1950s and 1960s to being awarded a commanding two-thirds majority at both the bachelor's and PhD levels. A decade ago, I heard a prominent social psychologist talking about the old days, when she was one of the few women in graduate school. She was referred to openly as "Blondie" and frequently had to listen to conversations about how women were fundamentally ill-suited to scientific psychology and naturally made for nursing, teaching, and homemaking. Furthermore, these opinions were justified by theories of biological determinism—the same ones people frequently offer to explain why women will never attain the same level of proficiency in math and science as men.

Such talk about psychology would be inconceivable today. That psychology is a much richer, more interesting, and more relevant field with so many women making contributions is self-evident. Would physics, computer science, and engineering enjoy similar benefits if we increased the ranks of women? There's reason to think so. Many efforts are currently under way to encourage girls to enter science, including single-sex classrooms and schools and many programs that include methods of boosting comfort and interest in math and science classes by applying what we have learned in our psychology laboratories. It will be interesting to watch over the coming decades how the results of this grand experiment turn out, if women continue to close the gap with men in areas where they have been underrepresented.

Whatever happens with regard to the low numbers of women in science, it is clear to me that great improvement is possible. But my dream result is not equal numbers of men and women in all fields. Rather, my dream is Charles Murray's dream, which he so eloquently shared during the conference on the nature and nurture of women in science: that every individual receives an education that leads to an enjoyable career that fits his or her unique talents and interests. I have another dream: That our educational efforts yield a population that is more scientifically literate and intellectually curious, with an interest in and respect for scientific data—so that even if we do not achieve equal numbers of men and women in science, more of us will be able to distinguish scientific arguments from ideology. That would be real progress.

Notes

1. Gladwell 2007.
2. National Academy of Sciences et al. 2007.
3. Ceci and Williams 2006.
4. Fallows 2004.
5. For reviews, see, for example, Aronson and Aronson 2007; Gladwell 1999; Harris 1998.
6. Jencks and Phillips 1998.
7. For reviews, see Aronson and Steele 2005; Aronson and McGlone 2008; Steele et al. 2002.
8. Aronson et al. 1999.
9. For an excellent discussion of sex differences in spatial abilities, see Newcombe 2006, as well as chapters 1 and 2, above, and chapter 9, below.
10. Connellan et al. 2000; see also chapter 1, above.
11. Schmader and Johns 2003; Seibt and Förster 2004; Derks et al. 2008.
12. Marx and Roman 2002; Marx et al. 2005.
13. Massey and Fischer 2005; Brown and Pinel 2003.
14. Cohen et al. 2006.
15. Good et al. 2003.
16. Cohen et al. 1999.
17. Murray 2008.
18. Aronson 2004.
19. Reardon et al. 2009.
20. Davis et al. 2006.
21. For reviews see Aronson and Steele 2005; Aronson and McGlone 2008.
22. Walton and Spencer, in press.
23. Stricker 2006; Sackett et al. 2004.
24. Wax 2004 and 2007.
25. Wax 2003.
26. See, for example, Flynn 2007.
27. Ceci and Williams 2006.
28. National Center for Educational Statistics 2003.
29. Dee 2005.
30. Dee 2007.
31. Pedersen et al. 1978.
32. Krueger and Whitmore 2002.
33. Jencks and Phillips 1998.
34. Murray 2008. Murray further argues that "the high-stakes tests given in elementary and secondary school are expressly described as measures of what students have learned, not how smart they are," but presents no evidence beyond the experience of his own children. Reardon et al.'s (2009) research on the California high school exit exam should give him pause about this assumption.

35. Stricker and Ward 2004; Danaher and Crandall 2008.

36. Massey and Fischer 2005.

37. Wax 2007.

38. Sackett et al. 2004.

39. Wax 2007.

40. Jencks and Phillips 1998.

41. Wax has tended to focus primarily on this tiny minority in a sea of published studies that did not use the analysis of covariance correction. Ironically, she accuses stereotype threat researchers of "cherry-picking" to arrive at their desired conclusions.

42. Brown and Day 2006.

43. Wax 2004.

44. A particularly thorough and entertaining explication of how and why scientists, politicians, district attorneys—and the rest of us—cook the facts to reach desired conclusions can be found in Tavris and Aronson 2007.

45. National Science Foundation 2006.

References

Aronson, E., and J. Aronson. 2007. *The Social Animal*. 10th ed. New York: Worth/Freeman.

Aronson, J. 2004. The Threat of Stereotype. *Educational Leadership*. November.

Aronson, J., M. J. Lustina, C. Good, K. Keough, C. M. Steele, and J. Brown. 1999. When White Men Can't Do Math: Necessary and Sufficient Factors in Stereotype Threat. *Journal of Experimental Social Psychology* 35: 29–46.

Aronson, J., and M. McGlone. 2008. Stereotype and Social Identity Threat. In *The Handbook of Prejudice, Stereotyping, and Discrimination*, ed. T. Nelson. New York: Guilford.

Aronson, J., and C. M. Steele. 2005. Stereotypes and the Fragility of Human Competence, Motivation, and Self-Concept. In *Handbook of Competence & Motivation*, ed. C. Dweck and E. Elliot. New York: Guilford.

Brown, R. P., and E. A. Day. 2006. The Difference Isn't Black and White: Stereotype Threat and the Race Gap on Raven's Advanced Progressive Matrices. *Journal of Applied Psychology* 91: 979–85.

Brown, R. P., and E. C. Pinel. 2003. Stigma On My Mind: Individual Differences in the Experience of Stereotype Threat. *Journal of Experimental Social Psychology* 39: 626–33.

Ceci, S. J., and W. M. Williams, eds. 2006. *Why Aren't More Women in Science?* Washington, D.C.: American Psychological Association.

Cohen, G., J. Garcia, and A. Master. 2006. Reducing the Racial Achievement Gap: A Social-Psychological Intervention. *Science* 313: 1307–10.

Cohen, G., C. M. Steele, and L. D. Ross. 1999. The Mentor's Dilemma: Providing Critical Feedback across the Racial Divide. *Personality and Social Psychology Bulletin* 25: 1302–18.

Connellan, J., S. Baron-Cohen, S. Wheelwright, A. Batki, and J. Ahluwalia. 2000. Sex Differences in Human Neonatal Social Perception. *Infant Behavior and Development* 23: 113–18.

Danaher, K., and C. S. Crandall. 2008. Stereotype Threat in Applied Settings Reexamined. *Journal of Applied Social Psychology* 38: 1639–55.

Davis, C., J. Aronson, and M. Salinas. 2006. Shades of Threat: Racial Identity as a Moderator of Stereotype Threat. *Journal of Black Psychology* 32 (4): 399–417.

Dee, T. S. 2005. A Teacher Like Me: Does Race, Ethnicity or Gender Matter? *American Economic Review* 95 (2): 158–65.

Dee, T. S. 2007. Teachers and the Gender Gaps in Student Achievement. *Journal of Human Resources* 42 (3): 528–54.

Derks, B., M. Inzlicht, and S. Kang. 2008. The Neuroscience of Stigma and Stereotype Threat. *Group Processes and Intergroup Relations* 11 (2): 163–81.

Fallows, J. 2004. When George Meets John. *Atlantic*. August.

Flynn, J. 2007. *What is Intelligence? Beyond the Flynn Effect*. Cambridge: Cambridge University Press.

Gladwell, M. 1999. *The Tipping Point.* New York: Little Brown.

————. 2007. None of the Above. *New Yorker.* December 17.

Good, C., J. Aronson, and M. Inzlicht. 2003. Improving Adolescents' Standardized Test Performance: An Intervention to Reduce the Effects of Stereotype Threat. *Journal of Applied Developmental Psychology* 24: 645–62.

Harris, J. R. 1998. *The Nurture Assumption.* New York: Free Press.

Jencks, C., and M. Phillips, eds. 1998. *The Black–White Test Score Gap.* Washington, D.C.: Brookings Institution Press.

Krueger, A., and D. Whitmore. 2002. Would Smaller Classes Help Close the Black–White Achievement Gap? In *Bridging the Achievement Gap*, ed. J. Chubb and T. Loveless. Washington, D.C.: Brookings Institute Press.

Marx, D. M., and J. S. Roman. 2002. Female Role Models: Protecting Women's Math Performance. *Personality and Social Psychology Bulletin* 28 (9): 1183–93.

Marx, D. M., D. A. Stapel, and D. Muller. 2005. We Can Do It: The Interplay of Construal Orientation and Social Comparisons under Threat. *Journal of Personality and Social Psychology* 88: 432–46.

Massey, D. S., and M. J. Fischer. 2005. Stereotype Threat and Academic Performance: New Data from the National Survey of Freshmen. *The DuBois Review: Social Science Research on Race* 2: 45–68.

McGlone, M., and J. Aronson. 2006. Social Identity Salience and Stereotype Threat. *Journal of Applied Developmental Psychology* 27: 486–93.

Murray, C. 2008. The Age of Educational Romanticism. *New Criterion.* May.

National Academy of Sciences, National Academy of Engineering, and Institute of Medicine of the National Academies. 2007. *Beyond Bias and Barriers: Fulfilling the Potential of Women in Academic Science and Engineering.* Washington, D.C.: National Academies Press.

National Center for Education Statistics. 2003. *The Condition of Education.* Publication No. 2003-067. http://nces.ed.gov/pubs2003/2003067.pdf (accessed April 24, 2009).

National Science Foundation. 2006. *Women, Minorities, and Persons with Disabilities in Science and Engineering.* http://www.nsf.gov/statistics/wmpd (accessed June 25, 2006).

Newcombe, N. 2006. Taking Science Seriously. In *Why Aren't More Women in Science?* ed. S. Ceci and W. Williams. Washington, D.C.: American Psychological Association.

Pedersen, E., T. A. Faucher, and W. W. Eaton. 1978. A New Perspective on the Effects of First Grade Teachers on Children's Subsequent Status. *Harvard Educational Review* 48: 1–31.

Reardon, S., A. Atteberry, N. Arshan, and M. Kurlander. 2009. Effects of the California High School Exit Exam on Student Persistence, Achievement, and Graduation. Paper presented at the annual meeting of the American Educational Research Association, San Diego, April.

Sackett, P. R., C. M. Hardison, and M. J. Cullen. 2004. On Interpreting Stereotype Threat as Accounting for African American–White Differences on Cognitive Tests. *American Psychologist* 59: 7–13.

Schmader, T., and M. Johns. 2003. Converging Evidence that Stereotype Threat Reduces Working Memory Capacity. *Journal of Personality and Social Psychology* 85: 440–52.

Seibt, B., and J. Förster. 2004. Stereotype Threat and Performance: How Self-Stereotypes Influence Processing by Inducing Regulatory Foci. *Journal of Personality and Social Psychology* 87: 38–56

Steele, C. M., and J. Aronson. 1995. Stereotype Threat and the Intellectual Test Performance of African Americans. *Journal of Personality and Social Psychology* 69: 797–811.

Steele, C. M., S. Spencer, and J. Aronson. 2002. Contending with Group Image: The Psychology of Stereotype and Social Identity Threat. In *Advances in Experimental Social Psychology*, Vol. 37, ed. M. Zanna. San Diego: Academic Press.

Stricker, L. J. 2006. Stereotype Threat on Cognitive Tests. Paper presented at the meeting of the International Military Testing Association, Kingston, Ontario, Canada.

Stricker, L. J., and W. C. Ward. 2004. Stereotype Threat, Inquiring about Test Takers' Ethnicity and Gender, and Standardized Test Performance. *Journal of Applied Social Psychology* 34: 665–93.

Tavris, C., and E. Aronson. 2007. *Mistakes Were Made (But Not By Me): Why We Justify Foolish Beliefs, Bad Decisions, and Hurtful Acts.* New York: Harcourt.

Walton, G. M., and S. J. Spencer. In press. Latent Ability: Grades and Test Scores Systematically Underestimate the Intellectual Ability of Negatively Stereotyped Students. *Psychological Science.*

Wax, A. L. 2003. Letter to the editor. *New York Times.* July 3.

———. 2004. The Threat in the Air. *Wall Street Journal.* April 13.

———. 2007. Stereotype Threat: A Case of Overclaim Syndrome? Transcript of presentation at the American Enterprise Institute. October 1. http://www.aei.org/events/filter.all,eventID.1536/transcript.asp (accessed January 14, 2009).

6

Stereotype Threat: A Case of Overclaim Syndrome?

Amy L. Wax

In math and science careers, men outperform women. Although many factors have been cited for these differences, the phenomenon of stereotype threat (ST) looms large as a favored explanation for observed disparities. Stereotype threat is a term coined by Claude Steele and his colleagues to refer to a psychological influence on test performance that derives from social expectations. The theory of ST predicts that, when widely accepted stereotypes allege a group's intellectual inferiority, fears of confirming these stereotypes cause individuals in the group to underperform relative to their true ability and knowledge. Men have long been assumed to possess superior talents in traditionally masculine fields such as mathematics and science. As a result, it is claimed, women face ST when attempting to perform in these domains.[1]

ST was initially described in a study investigating the reasons for the poorer performance of blacks than whites on standardized tests of academic aptitude. In an influential 1995 study authored by Claude Steele and Joshua Aronson, elite black and white Stanford University students were given an experimental test of verbal ability. Half were told the test would assess "individual verbal ability," while the rest were told that the purpose of the test was to evaluate psychological factors related to test performance. The authors theorized that the first instruction would call to mind stereotypes about blacks' inferior ability and thus would elicit an ST response, whereas the

I thank Jonathan Klick for helpful suggestions. Jason Levine and Alvin Dong provided excellent research assistance. All errors are mine.

second instruction would not have that effect. When resulting scores were adjusted for students' precollege scores on the verbal portion of the SAT (SAT-V), black students given the first (ST-diagnostic) instruction were found to perform below expectation, while those in the second (nondiagnostic) group performed as well as expected; white students, however, performed equally well under both conditions.[2] The authors concluded that the "threat"-induced fear of confirming stereotypes about black intellectual inferiority had caused black students in the threat-diagnostic condition to perform poorly.

In the wake of the Steele and Aronson paper, hundreds of studies and published journal articles have appeared that purport to document an impact for ST on test performance in a range of situations. Researchers claim that ST can depress test performance among lower socioeconomic classes,[3] Latinos,[4] the elderly,[5] and even groups that are not traditionally stereotyped.[6] Most notably, there is now a large body of work reporting that women perform worse on tests of mathematical skill under ST conditions—that is, when confronted with the stereotype of women's inferiority in math.[7] All in all, the phenomenon of ST has been analyzed extensively for over a decade and is now included in many standard psychology textbooks. Typing "stereotype threat" in a Google search yields thousands of relevant sites, many of which are mainstream media sources. ST has been repeatedly cited by newspapers, reported on television, and discussed in a variety of intellectual and political circles.[8]

It is not hard to see why advocates of social equality have seized on ST findings. If ST effects dominate, other causes of group performance disparities can be discounted. So, for instance, the Steele-Aronson observation that black students' verbal test scores are depressed under ST conditions suggests that longstanding test score disparities between blacks and whites might be due simply to performance anxiety rather than to real differences by race in academic ability, aptitude, or learning. The ST results also point decisively to broad social influences—most notably, invidious stereotypes and widespread assumptions of black inferiority—as the source of observed race gaps on commonly administered standardized tests, thereby banishing the bugbear of innate differences. But even conceding nurture, rather than nature, as the root cause of underachievement, attributing performance gaps to stereotype threat points away from arduous, long-term reforms like reducing discrimination or increasing a group's skill level. ST research raises the hope that underperformance is a short-term, situational problem that is amenable to the "quick fix" of altering testing conditions or revising test instructions. The clear implication is that, if assumptions based on invidious stereotypes can be dispelled, the performance of lagging groups will dramatically improve, and test gaps will disappear.

The promise of an easy road to equality extends to gender. If women's situation-specific response to unjustified group generalizations is the source of observed gender gaps in scientific success, then other oft-cited factors—such as differences in ability, interests, drive, priorities, or temperament—can be discounted. ST research also promises a low-cost fix for women's underrepresentation in science. If the signals that cause women to achieve less can be dispelled, observed performance disparities will abate, and the accomplishments of men and women in scientific and quantitative fields will quickly equalize. In keeping with these observations, a psychologist writing in an American Psychological Association (APA) volume on women in science notes that

> the stereotype threat research carries two implications. First, if a simple manipulation of instructions can produce or eliminate gender difference in performance on a mathematics exam, the notion of fixed gender differences in math ability is called into serious question. Second, stereotype threat is a result of cultural factors—specifically gender stereotypes about female inferiority at mathematics—and thus provides evidence of socio-cultural influence on gender differences in mathematics performance.[9]

In the same vein, a report by the National Academy of Sciences, entitled *Beyond Bias and Barriers: Fulfilling the Potential of Women in Academic Science and Engineering*, regards ST results as confirming the conclusion that innate gender differences play essentially no role in observed patterns of scientific achievement and occupational success. Rather, states the report, gender differences are "strongly affected by cultural factors," which "can be eliminated by appropriate mitigation strategies, such as those used to reduce the effects of stereotype threat."[10]

This chapter is about whether ST explains observed differences in performance between men and women on standardized tests of quantitative skill, or in math and science careers more broadly. Is there reason to believe that ST is the sole, or even the primary, explanation for the underperformance of females relative to males in these domains? After examining the key studies to date, the chapter concludes there is no basis as yet for identifying ST as an important, significant, or substantial contributor to observed gender disparities in test scores, academic achievement, or professional success in scientific fields. ST research to date has never shown that ST accounts for more than a trivial portion of observed gender gaps and thus fails to rule out a dominant role for other sources of female underperformance.

This is not to deny that the phenomenon of ST exists, nor that statistically significant ST-type effects have been demonstrated in some contexts. Many studies indicate that testing environment can interfere with test performance, with some groups perhaps more sensitive to these effects than others. Nonetheless, the ST literature raises serious questions about the significance of these results. The issue at the heart of ST research is this: How important is ST in explaining disparities in group achievement observed in the real world? More specifically, to what extent can gender differences in test performance and overall accomplishment be attributed to ST effects, as opposed to other causes? Does stereotype threat account for all, most, some, or only a little of women's underperformance relative to men on quantitative standardized tests and in scientific fields? Put more precisely, what percentage of the observed male–female gap in, say, math SAT (SAT-M) scores can be attributed to stereotype threat? In particular, what portion of the gap between men and women of outstanding ability—that is, those at the right tail of the bell curve who can be expected to comprise the great majority of high-achieving scientists—is due to ST?[11] These questions have not yet been squarely asked or answered. Despite the plethora of ST research, no study has precisely measured the magnitude of ST's effect relative to other influences on women's science and math performance overall. No study has told us "How much?" Yet that information will radically affect society's approach to women's underrepresentation in scientific fields. Specifically, if ST is the main culprit behind performance disparities between men and women, then resources should be directed almost exclusively to altering test instructions, improving women's working conditions, and countering the social stereotypes of women's lack of talent or interest in science. But if ST accounts for but a small portion of gender outcome differences, then efforts directed at manipulating testing conditions, boosting women's self-concept, or fighting social stereotypes are unlikely to yield significant results. Attention and resources are best expended in other directions. Alternatively, if existing disparities express genuine differences in talents, life priorities, or preferences, gender gaps might prove relatively intractable to manipulation. The best strategy would then be to do little or nothing about gender disparities in science careers.

In addressing pivotal questions about ST's relative contribution to real-world patterns of gender performance, this chapter does not purport to take on all of the ST literature in detail. Nor is it meant to be an exhaustive, technical review of study results. Rather, it seeks to highlight certain patterns in the research that raise questions and concerns about its implications and the significance of the reported findings. For reasons already noted, the temptation to identify ST as the chief source of group performance differences is compelling. To borrow a phrase from another context, ST's powerful appeal gives rise to what has been dubbed "overclaim syndrome": the habit of ascribing greater weight to a body of scientific evidence than the data can bear.[12] It is, therefore, not surprising that, as Paul Sackett and his colleagues have shown, ST research has generated a number of sweeping and potentially misleading claims.[13] The goal of this chapter is to counter the temptation to overclaim syndrome as applied to gender by achieving a more balanced and measured view of the ST research results.

What are some of the problems with current research that leave open the question of how much ST contributes to observed gender disparities? First, there is the issue of relative magnitude: What is the size of the ST effect compared to the gender gap in performance overall, and to the gap observed in selective segments of the population? Second, what is the baseline yardstick for assessing ST effects? Do there exist reliable or objective measures of skill in math and science, impervious to ST, against which ST effects can be precisely gauged? Third is the question of the scope of ST's influence: Does ST operate as a "threat in the air?" Is it "out there" as a default condition, pervasively affecting women's performance in contexts routinely encountered in the real world? Is ST the ordinary and expected condition of test-taking—and, by extension, of doing science more generally—such that it can be assumed to undermine women's performance at all times and everywhere? Relatedly, does most research either support or assume that special interventions are needed to *dispel* ST (implying ST is pervasively "out there" in the background), or is it based on the premise that special interventions are required to *create* ST (implying ST is not ordinarily just "out there")? Fourth is the problem of cherry-picking: Can the theory of ST explain why women do as well as or better than men in some measures of math performance (for example, grades in high school or college courses) but less well on standardized tests and in professional settings? And fifth, is there a novel approach to study design that might correct the deficits in ST research by generating crucial, missing information about the magnitude of ST's influence and its contribution to observed group differences?

The chapter concludes with a final challenge to ST research: if, as hypothesized, ST operates selectively to depress women's performance in math and science fields, how can that observation be reconciled with the full range of gender performance disparities, including those unrelated to quantitative domains? For example, why are women writers far less prominent and productive than men, even though women are widely believed to possess relevant talents that are equal to or better than men's? And what do these patterns imply for the plausibility of ascribing achievement disparities to ST more generally?

Relative Magnitude

Why do women's achievements in math and science fields fall short of men's? Because these fields draw heavily on quantitative ability, the attention of those seeking to explain these differences has been drawn to a longstanding gender gap in performance on the math portion of the SAT. The average scores of men and women on the SAT-M are not currently far apart, but the sex differential at the right tail of the bell curve, although fluctuating from year to year and narrowing somewhat over time, has always been substantial. In 2006, for example, the ratio of men to women scoring between 750 and 800 on the SAT-M was about 2.6 to 1.[14] This means that about 3.33 percent of the male test-takers scored in this interval, as compared to 1.29 percent of the females. The disparities are even greater in the upper reaches of this range. For example, between five and ten times as many boys as girls receive near-perfect scores on the SAT-M test in samples of mathematically gifted adolescents.[15] Student talent searches conducted at Johns Hopkins University yield similar ratios.[16] Since the most productive scientists are likely to come largely from this exclusive cohort,[17] it is important to investigate the sources of these differentials. How much of this lopsided ratio on the SAT-M is due to stereo-type threat? If ST were eliminated, would the ratio change much or at all? Would it disappear? Unfortunately, current research fails to answer these questions.

To understand why, it is necessary to take a closer look at actual ST stud-ies. Most of the key research is performed on university students who are drawn from contrasting demographic groups (blacks and whites; male and females). The goal is to compare the test performance of students from each group under conditions designed to elicit stereotype threat and under cir-cumstances that are not threat-inducing. Since it is only feasible to test each student once, subjects from each population must in turn be divided into an experimental category (tested under "threat" conditions) and a con-trol category (tested under "non-threat" conditions), generating four separate subgroups overall. The goal is to conduct a four-way comparison, thus inves-tigating if any difference can be shown in women's and men's performances under ST versus non-ST conditions.

Demonstrating an ST effect thus requires comparing test scores generated by four distinct groups of students. The problem is that the student subjects participating in any given study may have different levels of math ability. Accordingly, the average ability of students in each study category could differ as well. Thus, any observed difference in average test scores among the four groups of subjects in a particular study could reflect differ-ences in ability rather than ST effects—or it could reflect some mixture of the two. And it is impossible to tell from the raw scores on an experimental test how much each factor contributes to observed patterns. For example, if the female "threat" subgroup scores worse than the female "control" subgroup, that could be because the study subjects in the first group are genuinely, on average, less able in math. Or it could be because the "threat" test condition depressed their scores. Likewise, if no such difference in performance is seen in men, that could be because men are not influenced by ST. But the same pattern of results would be observed if ST did, in fact, depress the perform-ance of the male "threat" subgroup, but the men in that subgroup happened to possess greater average ability than the male "non-threat" controls. Taking the test under threat might then bring the average "threat" group score down to the average control-group level, creating the illusory impression that men are not vulnerable to threat.

To see this point, consider the following example. A researcher solicits student volunteers for an ST study. As is commonplace with these protocols, she chooses an equal number of male and female subjects, yielding twenty volunteers of each sex. She then randomly divides each group of twenty students into two groups of ten, to be assigned respectively to the experimental and control conditions. The men and women in the experimental group are given a math test under a "threat" condition. Those in the control group take the test under a non-threat condition. Assume that the average precollege SAT-M scores of students in each group turn out to be as follows:

- male non-threat (control)—590

- female non-threat (control)—590

- male threat (experimental)—605

- female threat (experimental)—550

The researcher then finds that women score significantly lower than men on the experimental test administered under "threat," but do as well as men when threat is removed. The results also show that men score somewhat higher under a threat condition than all other test groups. Assuming for purposes of this example that scores on the SAT-M reflect genuine math ability, does this observed pattern demonstrate an ST effect? The background SAT-M scores reveal that this pattern should not necessarily be interpreted this way. Rather, the scores on the experimental test might simply reflect average ability differences among the study subject groups. And even if test subjects are drawn from a relatively rarefied population—as would be the case for students attending a selective university—significant differences in ability levels could still exist.[18]

It follows that, in order to isolate and demonstrate any ST effects on test performance, subjects with similar background ability must be compared. There are two ways to accomplish this. The first is through statistical methods, such as adjusting performance for some reliable indicator of skill. Many researchers adjust experimental test results based on subjects' background SAT scores. This technique was used in the seminal 1995 paper by Steele and Aronson examining race differences in verbal ability,[19] and is employed in a number of gender studies as well. Alternatively, researchers use various techniques to restrict test subjects' range of abilities more narrowly. This can be done more or less precisely. Some choose subjects who have obtained SAT scores within a particular interval. Others draw their study subjects from students enrolled in the same university course, or with the same course background, or with similar grades in particular courses. A number of gender studies take this tack.

All these methods omit key information critical to assessing the explanatory significance and policy implications of demonstrated ST effects. To see this, it is necessary to look more closely at actual research results. For their 1995 study of black and white Stanford undergraduates, Steele and Aronson solicited volunteers from the undergraduate population as a whole. They observed that, when their subjects' scores on an experimental verbal test were adjusted for the students' college entrance scores on the verbal portion of the SAT, the resulting adjusted scores were lower for blacks than whites under the designated threat condition (that is, when test-takers were expressly told the test would reflect verbal ability), but about the same when no threat was imposed. The authors interpreted the results as suggesting that, apart from any ability differences as reflected in SAT scores, ST conditions independently depress the test performance of black, but not white, students (see figure 6-1).

A similar method was adopted by Johns and collaborators in investigating ST's impact on women's math performance.[20] Their results showed that female students drawn from a college introductory statistics course performed worse than their male counterparts after hearing an experimental test described as "a math test" (which the researchers designated the diagnostic or threat condition), but just as well when expressly warned about the dangers of stereotype threat prior to taking the test (designated as the control or non-threat condition). The female test-takers also showed no shortfall in performance when informed that the test was designed to gauge general problem-solving skills (designated as a "teaching intervention," see figure 6-2). As with the Steele and Aronson study, scores on the experimental test were adjusted for each student's background SAT-M score so as to facilitate comparisons among the four distinct groups of subjects (male and female control, male and female experimental; see figure 6-2).[21] Once again, the study format was designed to isolate the effects of ST on test performance and to leave aside (by adjusting away) any performance differences among the subgroups that might be due to disparities in background ability.

FIGURE 6-1

STEREOTYPE THREAT AND THE INTELLECTUAL TEST
PERFORMANCE OF AFRICAN AMERICANS

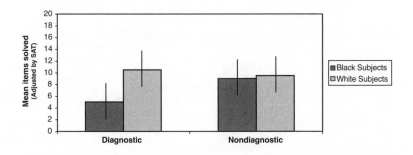

SOURCE: Steele, C. M. and J. Aronson. 1995. Stereotype threat and the intellectual test performance of African Americans, *Journal of Personality and Social Psychology* 69: 797–811.
NOTES: ST Condition (diagnostic instruction) = test problem solving ability; Non-ST Condition (non-diagnostic instruction) = determining psychological factors involved in solving verbal problems.

FIGURE 6-2

TEACHING STEREOTYPE THREAT AS A MEANS OF IMPROVING
WOMEN'S MATH PERFORMANCE

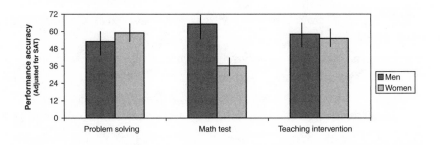

SOURCE: Johns, M., T. Schmader, and A. Martens. 2005. Knowing is half the battle: Teaching stereotype threat as a means of improving women's math performance. *Psychological Science* 16, 175–79.
NOTES: Women's and men's accuracy scores (adjusted for quantitative SAT scores) on the math test as a function of the test description. Error bars represent standard errors.

Other gender studies, rather than controlling for SAT differences directly, seek to match ability level more or less precisely through criteria for selecting research subjects. For example, in evaluating women's math performance under threat and non-threat conditions, Spencer and colleagues tested twenty-eight men and twenty-eight women drawn from a psychology class at the University of Michigan.[22] In the first part of their study, the authors confined their sample to students who scored in the 85th percentile on the SAT-M (above 650). When subjects were tested under a "threat" condition—in which they were told that the experimental math test was one that revealed gender differences—the women performed significantly worse than the men. When the subjects were instructed that the test produced no gender differences, the performances of men and women were comparable. Although the reported data were not statistically adjusted for SATs and other ability-related parameters, the authors asserted that a data reanalysis using these adjustments did not alter the results.[23] This suggests that the subjects in their admittedly "highly selected" sample of research volunteers were roughly "equally qualified,"[24] and, thus, that underlying ability differences across their research subgroups (male, female, experimental, control) were probably insignificant.[25]

What are the implications of studies like these? As noted, to distinguish score differences due to ST effects from those reflecting disparate underlying ability, researchers must either choose subjects with similar ability or adjust their subjects' performance scores for some background measure of individual skill. Although these methods have the merit of helping to distinguish effects due to ST from those due to ability, they also create costs. First, as stressed by Sackett and colleagues, controlling for background ability or restricting the skill range of study subjects can potentially mislead by creating the unwarranted impression that stereotype threat is the exclusive source of group disparities in performance among the study subjects and, by extension, in the population as a whole.[26] This impression, although not justified by the research results, can arise from the way the results are presented. For example, graphs that display test scores adjusted for background SATs will often show little or no difference in performance between the relative comparison groups (such as black and white, or male and female, test-takers), despite the fact that the study subjects themselves—and the broader populations from which they are drawn—may differ significantly in their ability levels as assessed by standardized test performance (see figure 6-1).

Likewise, studies that select subjects from a restricted range of ability levels can also create the misleading impression that *all* differences in test performance between groups (whether male–female or black–white) are due exclusively to stereotype threat. This impression arises from the fact that the range-restricted and ability-matched subjects in these studies are unlikely to represent an unbiased sample of the groups from which they are drawn. Because groups differ in their ability profiles, the degree to which a particular skill-restricted sample of subjects reflects the background population it represents will vary with each group. Indeed, in research designed to gauge ST effects by race or gender, study subjects matched for skill will almost certainly *not* be similarly representative of their background race or gender-specific population.

Consider a typical study designed to compare male and female math performance under stereotype threat conditions. Study subjects are chosen from students at a particular university. To qualify, all must have obtained a score of 750 or above on the math SAT. By definition, the men and women enrolled in the study will not be equally representative of the male and female populations as a whole. As noted, the ratio of men to women scoring above 750 on the SAT-M in 2006 was roughly 2.6 to 1. The male–female ratio toward the top of this range is even higher. Because women are significantly less likely to score above 750 than men, the female study subjects will be a more rarefied, and less typical, group than the men. In other words, the need to match the number and qualifications of study subjects across gender when investigating ST effects on women's math performance means that high-ability women in such research studies will be overrepresented, as compared to men, relative to their background same-sex population.[27]

The fact that men and women in typical ST studies are not likely to be similarly representative of their genders bears directly on whether these studies can answer the most critical questions: How big is the ST effect, and how much convergence in men's and women's scores can be expected from eliminating it? Consider once again the 2.6 to 1 ratio of male to female students scoring above 750 on the SAT-M. Would manipulations designed to dispel ST change that ratio significantly? Would altering test conditions elevate women's scores enough to match men's? The answer to that question depends on what portion of the gender differential is due to ST. And that in turn depends on the magnitude of the ST effect relative to gender disparities in math ability that are unrelated to ST.

The problem is that ST's relative contribution to the observed gender gap cannot be calculated by using commonly employed protocols for looking solely at matched cohorts of students at the right tail of the bell curve—or, for that matter, at any restricted portion of the skill distribution. Because there are significantly fewer women than men obtaining the highest scores, many women lower down on the curve would have to improve their scores significantly to close the gender gap at the top. More precisely, the gender gap would not disappear unless women all along the distribution scored higher, shifting women up the bell curve until their numbers equaled those of men at each interval. It follows, however, that if gender disparities in standardized math test scores are due largely to ST test anxiety, those anxieties must be assumed to depress the scores of women at all levels of performance.

The question of whether ST depresses women's real-world test scores all along the curve by a sufficient amount to account for existing gender gaps cannot be answered by the current crop of ST studies. That is because those studies consider small numbers of subjects over a restricted range. Even if male and female subjects in a relatively small test sample are observed to do better—or equally well—on an experimental math test under specified non-threat, as opposed to threat, conditions, it cannot be inferred that changing the SAT to make test conditions less "threatening"—or manipulating standardized testing instructions for the population as a whole—will close or even significantly narrow the male–female gap in math SAT scores at any particular achievement level. The effects currently observed in a small, unrepresentative slice of women tell us nothing about whether anything like the necessary improvement in female scores overall would occur if ST effects could be reduced. It is just as likely that most of the gap in actual background test results is due to "real" disparities in math aptitude or problem-solving ability—disparities that will not yield to short-term manipulations but, rather, are the product of other types of long-term influences.

In sum, the protocols commonly used in ST research, which control for background SAT scores or draw study subjects from a narrow ability range, leave crucial information on the cutting-room floor. By deliberately abstracting away from overall group differentials due to factors other than ST, these methods make it impossible to measure the magnitude of ST's contribution to score gaps relative to other causes. Because these studies provide no information about the comparative size of ST effects or the portion of existing background gaps that are due to ST, they tell us nothing about whether ST's influence is significant as compared to other factors like ability, knowledge, educational experience, interest in the subject matter, and learning.

For a concrete illustration of this problem, and of the potential for popular descriptions of ST research to mislead, consider a recent statement in a *New York Times* op-ed summarizing recent findings by Joshua Aronson and collaborators. The article states that "Mr. Aronson and others taught black and Hispanic junior high school students [that they] possessed the ability, if they worked hard, to make themselves smarter." According to the article, this intervention "erased up to half of the difference between minority and white achievement levels."[28]

The implication of this summary is that a large portion of the overall race achievement gap can be eliminated simply by telling students how capable they are. But this conclusion does not necessarily follow. We need to know far more about how this study was designed before leaping to such a dramatic conclusion. First, the op-ed report does not reveal whether the students in the study at issue were chosen randomly from the background population or whether they were matched for ability. Second, the description leaves us in the dark about the absolute magnitude of the ST-type effect observed relative to those subjects' test scores overall—or to any background achievement gap in the population as a whole. Indeed, we are told nothing at all about how the Aronson research was designed.

Consider one possible hypothetical scenario, which is fully consistent with the result reported in the op-ed piece. Suppose, as would be typical, that there is a significant disparity overall in the ratio of whites to minorities scoring in the top 10 percent on a standard junior high school achievement test. Suppose the ratio is 4 to 1, with whites even more dominant among the very top scorers. Suppose further that the subjects in the reported study were all selected to fall within that top 10 percent range. And assume, hypothetically again, that under high ST conditions, the white subjects in the Aronson study achieved an average score on the experimental test that was 5 percent higher than the minority subjects. Suppose also that under low ST conditions, that gap was reduced to a 2.5 percent average difference. Since 2.5 percent is half of 5 percent, the experiment can thus accurately be described as demonstrating a testing intervention that "erased . . . half of the difference between minority and white achievement levels."

However, because the students in the hypothesized studies were matched for academic ability, such a 50 percent reduction of the score gap from a manipulation in testing conditions would not be surprising. ST could be expected to account for a relatively large portion of the residual group difference in performance among study subjects with similar abilities. Yet that result is consistent with ST having only a small *absolute* impact on the scores of the relatively able minority students in the sample. Although reducing ST cuts that impact in half, the reduction is against the base rate of a very small absolute effect. A 50 percent reduction in a small number is a small number. Thus, the reported 50 percent gap reduction could be entirely consistent with an ST effect that is quite small relative to the (otherwise similar) scores of the matched study subjects.

The more important point, though, is that a study that compares selected white and minority students of similar ability tells us nothing about the ability profiles of the groups from which they are drawn. Those profiles reveal large group differences—differences that are necessarily masked by any study protocol that matches subjects for ability. Moreover, the magnitude of the ST effects observed in such a study could well be negligible compared to the size of these group differences overall. Certainly it does not follow from the study results stated in the op-ed that eliminating ST can reduce this *overall* minority–white performance gap by 50 percent or anything close to that.

How do these insights apply in the gender context? Would redesigning studies of male and female test performance to report scores adjusted for SATs in conjunction with unadjusted or raw scores solve the problem? No. Although presenting data in this way has the potential to provide more information about the precise portion of the score gap in a particular study sample that is due to ST effects (as opposed to ability differences), it does not reveal the relative size of ST effects in the population as a whole. The problem is again one of representativeness. As noted, there is no guarantee that subjects of any study, or any subgroup in any study, are typical of the background population or even of a defined segment of that population. Likewise, it cannot be assumed—and indeed, given current study designs and demographic realities, it is unlikely—that the comparison groups of subjects are equally representative, or typical, of their background populations.

This point applies to race as well as gender. In this vein, Steele and Aronson have recalculated their 1995 study results on ST effects on black and white students' verbal test performance using raw scores unadjusted for background SATs.[29] These numbers are summarized in an unpublished graph (see figure 6-3) provided by one of the authors.[30] When considered in conjunction with the SAT-adjusted data (see figure 6-1), the graph reveals that the black students in the study possessed lower average background verbal ability than the white students tested. It also shows that taking the test in the "threat" condition depressed black students' performance below the expected background levels, roughly doubling the preexisting racial performance gap.

Although examining both adjusted and unadjusted scores tells us something about the relative contribution of ST versus background skill to the black–white score gap in this particular study sample, it nonetheless fails to enlighten us on the contribution of ST to the black–white gap in SAT scores overall. Blacks scoring above 700 on the SAT-V are rare and much less common than white students scoring in this range.[31] Thus, the Stanford students tested for the Steele and Aronson study are not equally representative of blacks and whites as a whole, and may not even be similarly representative

FIGURE 6-3

VERBAL TEST PERFORMANCE UNCORRECTED BY SAT-V

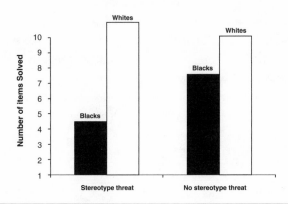

SOURCE: Unpublished graph from Joshua Aronson.

of students of each race who score in these students' elite range. For this reason, the fact that being tested under "threat" significantly depressed the performance of a small group of black Stanford students tells us nothing about the extent to which test manipulations could alter the overall ratio of blacks to whites with superior scores on the SAT-V test. Nor does it tell us the degree to which reducing ST could cause SAT scores for blacks and whites as a whole to converge. As with gender, narrowing the gap would require an upward shift in scores all along the ability distribution. Given the magnitude of the existing black–white SAT gap, that shift would have to be dramatic indeed.

The Skill Baseline

The majority of ST studies reported in the literature compare test perform-ance across distinct, non-overlapping groups of experimental subjects. The need presented by existing protocols to control for the skill level of study subjects poses the problem of how to measure real ability. Once again, an implicit assumption of many ST researchers is that "stereotype threat is an influence that may occur in an actual testing situation."[32] The implication is that the SAT-M gender gap—especially at the right tail—can be attributed mostly or exclusively to ST. But if ST does significantly depress SAT perform-ance, then the practice of adjusting for or limiting the range of subjects' SAT scores begs the question of whether the SAT provides an accurate baseline measure of ability independent of the ST effect that the studies seek to assess. This observation points to a potentially fatal contradiction in the design of much ST research: SAT scores cannot simultaneously represent an accurate measure of math ability, untainted by ST, while at the same time being vul-nerable to distortion by ST effects. If we accept that ST artificially depresses women's real-world test scores, then SATs do not reflect real math ability. Alternatively, if we posit that SATs are unaffected by ST, then ST effects can-not explain observed SAT gender gaps. Indeed, in that case, it is hard to see why we are interested in ST effects at all, since by hypothesis ST is irrelevant to the most important gender gap in real-world test scores!

In sum, ST researchers cannot have it both ways. They cannot use the SAT as an untainted, independent measure of ability and at the same time claim that ST explains some, most, or all observed gender differences in standardized math-test performance. This inconsistency in the ST literature has been noted more than once in the context of both gender and race,[33] and has never been satisfactorily resolved. Rather, it has generated a number of confusing and contradictory statements. One research group, for example, has defended its use of an SAT control for comparing high-scoring male and female math students by observing that "performance-depressing stereotype threat emerged in these studies only when the test was at the limits of [students'] skills." The authors went on to conclude that "it is very unlikely that stereotype threat hampered [the women subjects'] performance on the SAT exam they had taken just a few years earlier. It too was well within their skills, as indicated by their high scores." They added, nonetheless, that, "over the full range," the performance of at least "some" women on the SAT-M was "likely" to be affected.[34] The problem with this explanation is that the SAT gender gap is largest in the highest score range.[35] These authors are therefore suggesting that where score disparities are greatest, ST is least likely to explain them. The clear implication of this suggestion is that the SAT-M score gap at the right tail is not due to ST—but rather to real gender differences in math ability, whether innate or acquired.

In another paper, however, scientists from the same group imply that the women of highest ability are most vulnerable to ST effects,[36] while women who "dissociate themselves from math at an early age," and thus get lower scores on standardized tests, are least likely to respond to ST.[37] In short, the literature is rife with waffling on a number of critical issues, including whether commonplace tests of math ability are tainted by ST effects at all, whether ST is responsible for differential performance only in selected portions of the ability distribution, and which women at which skill level are most affected.[38]

It should be noted that an important piece of evidence appears to undermine the assertion that ST systematically distorts women's real-world performance on the SAT-M—and thus supports the position that the test is an untainted measure of baseline math ability. The hypothesis that ST is largely responsible for the SAT-M gender gap generates a particular prediction about test results. If ST artificially depresses women's background SAT scores, then men and women with matching SATs should not perform equally well under experimental conditions that eliminate stereotype threat. Rather, women should outperform men. Yet this pattern has not generally been observed.[39] These results cannot be squared with the claim that ST is an important source of group differentials in standardized test performance.

The Scope of ST's Influence

The issue of whether ST actually depresses performance on real-world tests is pertinent to yet another aspect of ST research, which is how ST experiments should be conducted. If real-world standardized tests are administered under conditions that are threatening to disfavored groups—so that observed score differentials can be largely attributed to ST—then it follows that ST is routinely present in ordinary testing situations. This means that ST is hovering out there "in the air" whenever anyone takes a test, so that no special measures or interventions are required to impose it. What are the implications of this assumption for experimental design? Because there is no need to create "threat," the administration of a test in the absence of any special instructions—or any instructions whatsoever—should constitute the diagnostic, experimental "threat" condition. In that case, however, creating the control, or non-threat, condition would appear to call for *affirmative* intervention. That is, the standing threat needs to be affirmatively removed or dispelled. Special instructions would therefore be needed to administer a test *without* the influence of ST.

Do social scientists consistently design their studies in keeping with these assumptions? Or do they implicitly assume that ST is not a pervasive background condition of all standardized testing, but rather taints test performance only in special circumstances? How do they generally define, identify, or create the experimental and control situations in ST research? How do they generate a "threat" testing condition, as opposed to a situation in which testing is free from threat? Once again, confusion reigns. Researchers in the field have not adopted a uniform protocol nor taken a consistent approach. In particular, the range of experimental designs reveals no consensus on whether ST is just out there "in the air," pervasively distorting the results of all standardized testing, or whether it is a condition that experimenters must create through special interventions or testing instructions.

To see this, consider the various ways researchers have generated threat and non-threat conditions. In one group of studies, scientists actively intervene to create the threat, usually by giving a specific pretest instruction. In this vein, researchers have told subjects about to take an experimental test that race[40] or gender[41] differences in test scores are to be expected. Or they have exposed subjects to gender-stereotypic television commercials prior to administering the test.[42] For the "control" or non-threat condition in these studies, in contrast, test-takers are either told nothing,[43] are given some kind of nongendered instruction (such as that the test is a gauge of personal math ability),[44] or are exposed to a stimulus (for instance, television commercials) with gender-neutral content.[45] These studies are generally most consistent with the implicit assumption that threat is not ordinarily "in the air," operating in most real-life test-taking situations, but rather must be specially created.

In contrast, other studies have researchers giving subjects special instructions for the purpose of dispelling or removing the threat. Thus, as reported in one paper, subjects in the control, or non-threat, group were told that the experimental test produced no gender differences and was "gender fair," while the "threat" (diagnostic) group was told nothing at all about gender.[46] In another study, the goal was to investigate "whether reminding women of other women's achievements might *alleviate* women's mathematics stereotype threat."[47] Thus, women who were about to take a difficult math test were informed that women make better psychology study subjects than men, or were read profiles of accomplished professional women. The expectation was that this group's performance would be unaffected by ST—that is, these instructions were supposed to generate a non-threat or control condition. In contrast, the "threat" group—which was expected to and did achieve lower scores—was given a gender-neutral reading about successful corporations. In yet another study, college-age mentors encouraged seventh-grade female subjects "either to view intelligence as malleable" or to ascribe their academic difficulties "to the novelty of the educational setting."[48] These student subjects delivered a better test performance than other girls who were given no such instructions. Studies of this type are more consistent with the assumption that all tests are taken "under threat," regardless of testing instructions. It follows that ST will operate to depress vulnerable groups' real-world performance unless specific steps are taken to blunt or remove its influence.

In still other articles, researchers used specific test instructions both to create and to dispel threat. In one, for example, some subjects were told that women were expected to do worse on the experimental test, while others were told that men and women performed equally well.[49] Yet other studies adopted a range of manipulations for comparing performance under supposed ST and non-ST conditions, including administering a test in mixed-sex or single-sex groups,[50] telling some women the test was designed to expose intellectual strengths while informing others that it highlighted intellectual weaknesses,[51] testing subjects in the threat and non-threat conditions in the presence of background noise while instructing some that the noise would likely depress their scores (that is, giving a so-called misattribution instruction in conjunction with an ST or gender-neutral condition),[52] and coaxing women into thinking more generally about their strengths rather than their stereotypical weaknesses.[53]

The dizzying array of research protocols raises obvious questions about the assumptions that inform these study designs. Specifically, when, if ever, must ST effects be affirmatively generated, and when must they be dispelled? What is the theory behind the answers to these questions, and what is the implication for whether and when ST operates on real-world testing? Can the so-called "threat" conditions in ST studies be analogized with real-life testing conditions? Are the study protocols consistent with the assumption that most testing—including math SAT testing—is conducted under "threat," or do they assume that most testing is free from threat? In other words, is there sometimes, often, or always a residual background ST effect "in the air"? Does threat require a special intervention—say, in the form of a gender-salient test instruction—or is it just "there" as the normal condition under which tests are generally taken, so as to require no special instruction? Why do some experiments show women performing as well as similarly skilled men when they are given no instruction (but underperforming after a threat-enhancing instruction), whereas others show women performing worse with no instruction (but performing just as well with a threat-dispelling instruction)? Is there an inconsistency here? One searches in vain for any analysis of these issues. Indeed, there is little systematic discussion in the ST literature of how theoretical expectations should inform research design, and virtually no consideration of whether the ST data as a whole are well-behaved in light of theory.

Cherry-Picking: The Selective Operation of ST

Yet further anomalies in the literature raise questions about the operation of stereotype threat within various domains that call upon math and science skills. Specifically, can ST explain the uneven pattern of female participation and achievement in these areas overall? Women are now about as likely as men to take advanced quantitative courses in high school and to major in math and science fields as undergraduates. That women earn better grades than men in high school and college math courses is often cited as evidence of their equal ability in these areas.[54] Yet women's enrollment in graduate school, their rates of professional advancement, and their productivity as working scientists lag behind.[55] Why does ST not diminish women's performance in the classroom or on class-related tests? Why are women not worried about confirming stereotypes in these contexts? The influence of ST would be expected here, especially in light of studies suggesting that mixed-sex settings (like coeducational college and university classes) generate ST threat effects and inhibit performance.[56] The few explanations offered—that, for example, standardized tests are generally intellectually demanding whereas coursework is uniformly "well within [women's] ability," or that women's experience of success within the classroom helps dispel stereotype threat[57]—are either questionable as a matter of fact (since upper-level math courses can be quite challenging) or circular (since women's record of classroom success just begs the question of why ST does not undermine that success in the first place). In short, attempts to account for observed patterns are, as yet, unsatisfactory.

Additional questions remain. Are ST effects cumulative and additive, or do they conform to an on–off pattern, such that someone either experiences the threat (with a fixed effect of determinate size), or not? If ST is "in the air," can researchers nonetheless further depress women's performance by giving a specific threat-generating instruction? Are ST effects on test performance linear in their impact—that is, do they sum up in a straightforward way? Different answers to these questions predict different results for ST research. The failure to match up theory to results—to come up with more precise hypotheses about how ST operates and then to devise studies designed specifically to confirm or disconfirm—is a serious flaw in the literature. These omissions represent yet another way in which social scientists have ignored important quantitative dimensions of ST.

Although ST research has so far been directed at validating the existence of the phenomenon, the next stage should undertake a more precise calibration of ST's magnitude relative to other influences on outcomes for men and women. The failure systematically and precisely to measure ST's impact over the full range of conditions makes it impossible to determine the size of ST effects as compared to other factors that can produce gender or group differences in performance. Yet knowledge of this relative magnitude is absolutely essential to an accurate assessment of ST's significance, which in turn is necessary to the development of a scientifically informed, rational strategy for dealing with differential group achievement. In particular, quantitative information is essential to any action plan for addressing gender gaps in math and science performance.

ST Study Design: Answering the Unanswered Questions

What questions should ST researchers now seek to answer? Put baldly, does ST account for 1 percent, 10 percent, 50 percent, 80 percent, or all of the gender difference in performance on standardized tests of math and science aptitude? What portion of the gender gap at the right tail of the bell curve—and in the number and achievements of the most productive scientists—can be attributed to ST? Addressing these questions requires measuring the background, real-world influence of ST, fixing a reliable baseline for its measurement, and gauging its relative contribution to existing disparities. These tasks cannot be accomplished without a paradigm shift in ST research. In particular, determining *how much* ST contributes to observed gender disparities calls for a radical new approach to ST study design.

How might ST research be structured to reveal the pertinent information? More generally, is it possible to create a research protocol to address the key unanswered questions: Does ST account for all, some, or only a little of the gender gap in scores on standardized tests like the SAT-M (the relative magnitude problem)? Does ST significantly depress women's scores on standardized tests such as the SAT-M, or do such tests represent an accurate, untainted measure of real mathematical acumen (the baseline problem)?

One option is to begin with a well-defined working hypothesis. Although there is much equivocation on this point, assertions in the pertinent literature—

such as the statement quoted earlier from the APA volume addressing women's underrepresentation in the sciences[58]—strongly suggest that ST is a major, if not the sole, source of the gender gap in math and science performance. Therefore, one possible initial hypothesis is this: The gender gap in the SAT-M is due exclusively to ST. But if, in keeping with this hypothesis, it is assumed that "stereotype threat is responsible for the underperformance of women in quantitative domains," then it follows that "removing stereotype threat from those situations should eliminate women's performance deficit."[59]

How could this prediction be tested? That is, how could it be shown that eliminating ST's influence would close *all*—as opposed to some or none—of the gender gap in math and science performance? One possibility is to focus, as many gender studies already do, on a particular slice of the test-taking population—but to take a different approach. The women most likely to become prominent scientists are the ones at the extreme right tail of the bell curve— that is, women who score 750 and above on the SAT-M. As already noted, in 2006, 3.33 percent of male SAT test-takers scored between 750 and 800, while only 1.29 percent of female test-takers did so.[60] For purposes of illustration, a possible distribution of men's and women's scores consistent with these ratios is schematically depicted in figure 6-4.

FIGURE 6-4

MATH SAT: "NORMAL" (STEREOTYPE THREAT) CONDITIONS

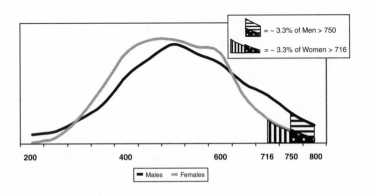

Now, in keeping with our hypothesis, assume that men and women do "truly" possess equal math ability, and that the entire gender disparity for top scorers results from the operation of ST. A corollary of these assumptions is that, if stereotype threat could somehow be entirely dispelled, the percentage of women and men scoring 750 and above would precisely equalize. This means that the percentage of women test-takers scoring in this range would rise to 3.33 percent.[61] Accordingly, the distribution of men and women at the right tail of the bell curve would be the same. Indeed, the consequences of our hypothesis can be summarized more broadly: If the gender gap in SAT-M scores all along the distribution—including at the right tail—is due entirely to ST, then removing the influence of ST should cause the bell curves for male and female SAT-M performance to converge. That is, the percentage of males and females achieving each score would be equal. This result is schematically depicted in figure 6-5.

FIGURE 6-5

**MATH SAT: STEREOTYPE THREAT REMOVED
(ASSUMING ST CAUSES THE GENDER GAP)**

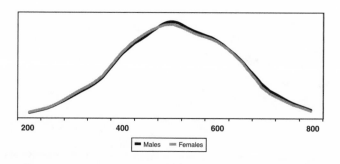

| 200 | 400 | 600 | 800 |

■ Males ■ Females

SOURCE: Author's illustration.
NOTE: Figure assumes ST causes the gender gap.

A comparison of the actual distribution of SAT-M scores (as reflected in figure 6-4) and the distribution (as reflected in figure 6-5) that would be predicted to result, in our hypothesis, from the removal of ST (if indeed ST is the sole cause of gender score disparities) makes it possible actually to measure the precise magnitude of ST's effect on women's SAT-M performance. The key is to focus on the following question: Given the existing profile of SAT-M scores by gender, what is the score above which the percentage of women is equal to the percentage of men scoring 750 or higher? That is, what is the lowest score women would currently have to achieve to be in a group of equal relative size to that for men scoring at or above 750? Since 3.33 percent of men are in this range, we look for the minimum score actually achieved by the same percentage (3.33 percent) of women, which is roughly 716. Accepting our hypothesis, this allows us to estimate that ST depresses women's SAT-M scores, at least in this part of the ability distribution, by approximately thirty-four points.

This information is critical to determining whether our hypothesis is correct, because it permits us to decide whether ST in fact accounts for all, or some smaller part, of the gender disparity in SAT-M performance. Our hypothesis predicts that, if women scoring 716 or above on the math SAT could be retested without the influence of ST, their scores would significantly increase. More precisely, if ST is the sole cause of the gender gap, the scores of this cohort of women should rise to match men's—that is, to 750 and above.

How would we conduct this experiment? Ideally, it would be possible to identify women scoring above 716, and to select a cohort from this group that would reflect the distribution of women in this range; likewise for men scoring 750 or above.[62] Half of these men and women would then be asked to take an experimental math test under threat, and the other half in a non-threat condition. The performance of the men and women would then be compared. (The study's hypothesis is that the SATs are tainted by threat, which implies that threat is always out there "in the air." Consistent with this, the ST threat condition should involve administering the test with no special instruction, and the non-threat, or "control," condition would involve an instruction to dispel or eliminate the threat.)

What results would our hypothesis predict? In the threat—that is, normal testing—condition, the experimental test should show a gender gap that reflects the background gap in SAT scores for the study subjects. But administering the test under conditions that dissipate the threat should cause the gender gap to disappear. That is, the women in the study sample should achieve the same scores—on the same distribution—as the men. The bell curves in the subject groups should converge. In sum, if ST is the only reason for the observed SAT score gap, the male and female study subjects should, on average, achieve the same profile of scores in the non-threat condition, despite women's lower background SAT scores. This reflects the understanding that, in the absence of threat, the same percentage of women as men will achieve each score.

Suppose that this result is not observed? It follows that our strong initial hypothesis—that ST is responsible for the entire SAT-M gender gap—is false. The experiment is nonetheless informative. Suppose, for example, that men's and women's scores narrow somewhat in the non-ST condition. Then measuring the extent of remaining divergence will allow a precise "decomposition" of factors responsible for the gender gap. Specifically, quantifying the remaining degree of divergence would enable researchers to measure exactly how much women's SAT-M scores are actually depressed by ST effects and how much of the gap is due to other influences. This would permit an assessment of the magnitude of ST's impact on women's SAT performance relative to other factors. This is the information that is currently missing—and just precisely what we are seeking.

The degree of gender-score convergence observed in this experiment also tells us something about the SAT as a baseline yardstick of "real" math ability. Indeed, if male and female scores in our experiment are observed to converge slightly or not at all, there is good news and bad. The good news is that the SATs look to be a true and objective measure of ability, unaffected by ST effects. Researchers would therefore be justified in adjusting experimental test results for background SATs as a way to compare subjects of unequal ability and to isolate the influence of ST. But the bad news is this: If the SATs are, indeed, a true and objective measure of ability untainted by ST effects, then it follows that ST can't be the source of the gender gap in SAT-M performance. That is, ST can't explain women's underperformance on these tests. But that begs the question of why we should care about ST effects at all. By definition, ST has little influence on the most important—and powerfully predictive—assessment of aptitude for math. It follows that the real reason for women's underperformance must lie elsewhere.

ST and the Problem of Pervasive Disparity

A final caveat on ST research is in order. In touting the influence of ST on women, social scientists have focused almost exclusively on performance in selected areas—in particular, math and science. Because hoary stereotypes and traditional expectations about women's talents and interests have long held sway in these fields, the belief that pervasive cultural stereotypes impede women's performance in these arenas is widespread.

The problem with this selective focus is that women's underrepresentation in positions of achievement and influence is not confined to quantitative and scientific careers. Rather, men outperform women across the board, with women relatively scarce at the top of fields drawing on a broad range of aptitudes, including those for which women equal or outperform men on standardized tests and other well-accepted measures of ability. Dramatic gender disparities in achievement, productivity, output, occupational participation, and prominence persist even in areas where cultural beliefs regarding women's inferiority are absent, or where gender differences in ability have not been demonstrated, at least by conventional metrics.

Consider magazine writing, book authorship, and journalism. These endeavors require proficiency in writing and reading literacy—areas in which women are widely thought to excel and consistently outscore men on standardized tests.[63] Whether there are or ever will be equal numbers of men and women with the highest ability in math and science has been subject to vigorous debate, but few have suggested that women fall short of men in verbal skills. In light of these observations, the influence of gender stereotyping—and gender-based ST—is not generally believed to depress women's performance in these areas. Indeed, that women's achievement drawing on verbal abilities is unaffected by ST is an oft-stated assumption behind ST research designed to demonstrate the selective influence of ST on women's math and science performance.[64]

Yet women's "natural" verbal skills have not translated into dominance of fields drawing on these abilities. In particular, girls' strength in writing at all educational levels is not reflected in women's relative success in journalism or productivity in authorship of books and magazine articles. Among the books designated by the *New York Times* as the ten best of 2007, only two were written by women.[65] Of the thirty additional books recommended by the editors of the *New York Times* for 2007, seven were by women authors. In addition, the thirty-one winners of the 2008 Pulitzer Prize for writing and reporting included seven women.[66] Likewise, a routine perusal of advertisements by prominent publishing houses and university presses reveals a consistent and pronounced predominance of male authors. A tally from recent publication lists confirms this impression. For books released by a sampling of scholarly publishers between January 2007 and March 2008 in history, philosophy, the social sciences (sociology, political science, psychology, economics, anthropology), public policy, and literature, men strongly out-number women authors in all fields except literature.[67] Finally, an informal survey of pieces published in leading journals of opinion over the past three years reveals a decidedly lopsided pattern of authorship across the board, with male to female ratios of 28 to 1 for *Foreign Affairs*, 6 to 1 for the *New York Review of Books*, 7 to 1 for the *New Republic*, 6 to 1 for the *Atlantic Monthly*, and 4 to 1 for the *New Yorker*.

Can ST explain these dramatic disparities? Unlikely. But the persistence of wide gaps in productivity and achievement in areas conceded to be unaf-fected by ST casts doubt on ST's importance in math and science fields as well. Although the mix of factors leading to gender gaps need not be the same in all domains, the principle of Occam's razor suggests that those who would posit very different mechanisms for female underrepresentation across diverse fields bear the burden of persuasion. Given male dominance in occupations across the board—including many for which women are not stereotyped as less capable—it is important to step back and consider whether ST really accounts for most observed gender disparities. Factors that apply more broadly to many different endeavors should receive due consideration.

What unifying explanations can be offered? Perhaps authorship is not just a matter of verbal facility. Intellectual attributes of a more general kind, as measured by instruments such as IQ tests, may also be implicated. Although women and men are equal in average IQ, men outnumber women on the tails of the IQ distribution, with more men achieving the very high-est scores.[68] It is far more likely, however, that women's relative lack of prominence is traceable to average gender differences in temperamental or "conative" traits such as competitiveness, ambition, singlemindedness, and drive,[69] or to women's greater attraction to and interest in people rather than things,[70] or to other gender disparities in patterns of intellectual interest,[71] focus on career advancement at the expense of domestic pursuits, or desire to achieve life balance.[72] In short, available evidence suggests that ST explains relatively little of the patterns of male and female accomplishment observed in the real world today. Surely something else is going on. Although continued investigation of ST is certainly warranted, exaggerated claims for ST's significance should be avoided. A clear-eyed assessment of all the evidence is the only cure for overclaim syndrome.

Notes

1. See, for example, Davies and Spencer 2004, 174.

2. Steele and Aronson 1995, 799.

3. Croizet and Clair 1998.

4. Aronson and Salinas 1997.

5. Levy 1996.

6. Aronson et al. 1999.

7. See, for example, Davies and Spencer 2004.

8. See, for example, Sackett et al. 2004b. See also Sackett et al. 2004a and Sackett et al. 2005.

9. Hyde 2006, 138–39.

10. National Academy of Sciences et al. 2007, 49.

11. Studies of the correlation between standardized math test performance by gifted students and subsequent careers in math and science show that those scoring in the top 0.25 percent (top 1 in 400) exceed even those in the top 0.75 percent on various metrics of success, such as tenured positions, pay, age at promotion, and so on. These data suggest that high performance on standardized math tests is a significant predictor of future scientific accomplishment and career success. See, for example, Lubinski and Benbow 2006.

12. See Morse 2006.

13. See, for example, Sackett et al. 2004b, 8, documenting statements in the media and by the executive director of the American Psychological Association indicating that removing stereotype threat would have the effect of equalizing the verbal and standardized test scores of blacks and whites.

14. College Board 2006. See also Ceci et al. 2009 (discussing the significance of gender score differentials on "gatekeeper" tests such as the SAT-M).

15. See data cited in Murray 2005.

16. Halpern et al. 2007, 12–15. See also Ceci and Williams 2006, 147–57.

17. See Park et al. 2008, 957–61, showing that, even among students scoring in the top 1 percent of the math SAT at age thirteen, those scoring in the top quartile of that elite group were significantly more productive and successful in future scientific work than those scoring in the bottom quartile.

18. Specifically, there is no reason to expect that ability profiles for students at selective colleges will be identical by race or gender.

19. Steele and Aronson 1995, 799.

20. Johns et al. 2005.

21. Like Steele and Aronson (1995), the authors of this study adjusted scores for background SAT-Ms globally—that is, both within and across gender—which had the effect of abstracting away from any male–female differences in math ability that might have existed in the study sample. In contrast, Brown and Josephs, in a different gender ST study (1999), used the alternative technique of adjusting experimental test scores for

background SAT-Ms only within each gender. This method allowed the comparison of women's scores under ST and non-ST conditions, respectively, by controlling for any ability differences between the groups of women tested under these two conditions. But it left uncontrolled any ability differences across gender. The Brown and Josephs study design purported to respond to a particular criticism leveled at the Steele and Aronson study protocol, which is that comparing experimental scores adjusted for SATs across groups with very different background SAT profiles may create the unwarranted impression that observed differences in group performance—whether for study subjects or for the broader population from which the subjects are drawn—are due *entirely* to stereotype threat. With respect to blacks and whites, that impression is clearly misleading. See Sackett et al. 2004a and 2004b, and Wax 2004. See also discussion and note 25, below. Although the Brown and Josephs study design eliminates this particular methodological flaw, it still fails to provide any information on the size of the observed ST effect relative to the male–female test gap overall. And, as explained more extensively later in the chapter, the study thus provides no further reason to conclude that stereotype threat is a significant source of the gender gap in math standardized test performance. Indeed, it contains no useful data at all on this question.

22. Spencer et al. 1999.

23. Ibid., 9n2.

24. See ibid., 13–14.

25. The authors repeated their experiment for students with math SAT scores between 400 and 650. This time, they designated as the "threat" condition a test instruction that made no mention of gender at all. The non-threat condition was created by instructing test-takers that the test generated no gender differences in results. The women in the threat condition were observed to underperform relative to the male students. No significant performance differences were measured for students in the non-threat condition. See Spencer et al. 1999.

26. See, for example, Sackett et al. 2004a and 2004b, and Wax 2004.

27. Biased sampling by gender for research seeking to compare cognitive attributes of men and women is a problem that can cut both ways. For example, studies of profoundly gifted early adolescents have long shown a pronounced male advantage in mathematics aptitude, with male to female ratios for extremely high scores as large as 13 to 1. Yet most of the data come from the 1980s, when girls were probably less likely than boys to be identified and referred for a study of math and science talent. That the pronounced male advantage has strongly moderated recently suggests that girls were underrepresented in the early years of these studies relative to the number of mathematically gifted girls in the population. See Halpern et al. 2007 and Ceci and Williams 2006.

28. Nisbett 2009.

29. See Steele and Aronson 1995.

30. The graph (which also appears in chapter 5, above), sent by Joshua Aronson, was not accompanied by information on number of students in the study, raw SAT

scores, precise study design depicted, or significance level of the data. Aronson did not supply these data despite repeated requests.

31. In 2006, the ratio of whites to blacks scoring above 700 on the verbal SAT was 7 to 1, with 5.4 percent of all white test-takers scoring in this range, as compared to 0.74 percent of all black test-takers. On the math SAT, only 0.6 percent of all black test-takers scored at least 700, compared to 6.4 percent of all white test-takers. Thus, whites were more than ten times as likely as blacks to score 700 or above on the math SAT. Overall, the average difference between black and white scores for the combined math and verbal portions of the SATs is about 200 points, and it has remained steady for a decade. See *Journal of Blacks in Higher Education* 2006. See also Jencks and Phillips 1998.

32. See Hyde 2006, 138–39.

33. See, for example, Jencks and Phillips 1998, 424 n38: "If stereotype threat significantly undermines the performance of black students on the SAT, is it appropriate to use the SAT to equate the skill levels of black and white participants in our experiments?" See also Spencer et al. 1999, 25: "If stereotype threat depresses women's performance on standardized math tests relative to that of men, one might ask whether it is appropriate to use the SAT as a means of equating men and women for skill level in these experiments."

34. See Spencer et al. 1999, 25n10.

35. See Quinn and Spencer 2001, 57. See also Jencks and Phillips 1998.

36. See Quinn and Spencer 2001, 57: "Why are women who have achieved the most, who have the strongest math skills, underperforming in comparison to their male peers? . . . We believe the answers to these questions lie in examining the interaction between cultural stereotypes and the test-taking situation, what we call a 'stereotype threat' situation."

37. Ibid., 59.

38. For more discussion on the conflicting currents in the research on this point, see Sackett et al. 2004b. In any event, ST is unlikely to be an important contributor to the SAT-M score gap if it depresses test performance only for women at selected skill levels. As noted, closing the gender gap in the math SAT would almost surely require women to improve their scores in all portions of the ability distribution.

39. See, for example, Johns et al. 2005, Spencer et al. 1999, and Quinn and Spencer 2001, failing to show systematically higher female performance under non-threat conditions for an SAT matched sample. A similar pattern is observed for race and carries the same implications. See Jencks and Phillips 1998, 424, observing that "if stereotype threat lowers the performance of able black students on the SAT but not in our nondiagnostic [that is, non-ST-influenced] group, blacks in the nondiagnostic groups should perform *better* than whites with the same SAT scores," but noting also that this predicted pattern "rarely occurs."

40. See, for example, Steele and Aronson 1995, 799, the *locus classicus*, in which the diagnostic (ST-threat) group was instructed that the test would "provide a genuine test of your verbal abilities" but the control (non-ST threat) group was told the test's

purpose was to investigate the psychology of test-taking generally, with no mention of verbal ability.

41. See, for example, Spencer et al. 1999, study 1, and Schmader 2002.

42. See Davies et al. 2002.

43. See Spencer et al. 1999, study 1.

44. See Schmader 2002.

45. See Davies et al. 2002.

46. See, for example, Spencer et al. 1999, study 3; Quinn and Spencer 2001; and Ben-Zeev et al. 2005.

47. McIntyre et al. 2003; emphasis added.

48. Good et al. 2003.

49. See, for example, Spencer et al. 1999, study 2, and O'Brien and Crandall 2003.

50. See, for example, Ben-Zeev et al. 2005; Inzlicht and Ben-Zeev 2000; and Inzlicht and Ben-Zeev 2003.

51. Brown and Josephs 1999.

52. Ben-Zeev et al. 2005, study 2.

53. McGlone and Aronson 2006.

54. See Spelke and Grace 2006, 60, and Davies and Spencer 2004, 176, noting that the gender gap on standardized math tests "is not replicated in classroom grades." See also Ceci et al. 2009, 244, noting that "[o]ne puzzling aspect of the stereotype threat findings is why the anxiety from self-evaluative threat in the presence of implicit negative stereotypes has not resulted in female students actually learning less mathematics than male students or doing poorly on school achievement tests."

55. National Academy of Sciences et al. 2007, 59–85.

56. See, for example, Ben-Zeev et al. 2005; Inzlicht and Ben-Zeev 2000.

57. See, for example, Davies and Spencer 2004, 176.

58. See note 9.

59. See, for example, Davies and Spencer 2004.

60. See discussion earlier in the chapter.

61. This statement rests on the assumption that men and women who score in the upper reaches of the SAT are roughly equally representative of the same-age male and female populations as a whole. There is good reason to believe this assumption is sound. A substantial percentage of high school students now take the SATs. In light of this, and given the single-digit percentages of students achieving top scores, it is not implausible to assume that virtually all students, whether male or female, who are capable of achieving scores in the 700 range now sit for the test. Therefore, it is highly unlikely that female test-takers in this group are more skewed either toward lower or higher ability than males, or vice versa. In other words, men and women test-takers in this cohort are almost certainly equally representative of the ability-profile of their gender as a whole.

62. Although it might not be easy to find such a precisely matched sample, it should not be impossible. A large, elite university should be able to generate groups of study

subjects of reasonable size that mirror the distribution of male and female SAT-takers in the top tier. Alternatively, researchers could pool results from several universities to achieve the requisite number of data points. Or subjects drawn from some narrower interval of scores could be compared, so long as subjects were selected to match the respective percentages of men and women who actually achieve particular scores. For example, the percentile profile of women scoring from 716 to 730 might mirror that of men scoring from 750 to 760. Determining the precise way in which various male and female subjects should be selected would require calculations similar to those performed above. Once again, these calculations would be based on the assumption that all disparities in performance—and any differences in bell curve distribution—are entirely due to ST. This means that, without ST, the same percentages of women and men should achieve the same scores.

63. Although men have historically slightly outperformed women on the SAT-V and other timed tests of verbal ability that stress vocabulary, analogies, and logic, women usually score higher in assessments of writing and critical reading. See, for example, Kelley 2007, A11, noting girls' higher average scores in the recently added SAT writing component; see also Halpern et al. 2007, 6–7, noting females' "consistent outperformance of males in writing achievement" and in "reading literacy."

64. See, for example, Spencer et al. 1999, 22: "The stereotype about women is relatively confined—pertaining mainly to math and science"; Davies et al. 2002, 1621: "Because cultural stereotypes do not accuse women of having inferior verbal skills, women in verbal domains do not risk being personally reduced to negative stereotypes"; and Cullen et al. 2004, 225: "There is no basis we are aware of for positing that women systematically experience threat when taking the SAT [verbal]."

65. See New York Times 2007.

66. See Perez-Pena 2008, B7.

67. The surveyed publishers were the university presses of Oxford, Cambridge, Harvard, Yale, and Johns Hopkins, as well as the Sage Press. The ratios of identifiable male to female authors (including for multiauthored and edited volumes) were: social sciences, 3 to 1; public policy, 1.7 to 1; philosophy, 4 to 1; history, 2.5 to 1; literature, 1 to 1.

68. See, for example, Lubinski and Benbow 2006, 94–95

69. Ibid., 88–93.

70. See, for example, Baron-Cohen 2003.

71. Lubinski and Benbow 2006, 87–88.

72. Ibid., 91–92. See Ceci et. al 2009 (suggesting that tastes and preferences are the primary drivers of gender disparities in math and science career success). Old-fashioned discrimination is always a possibility, but is implausible in most contexts discussed in this section. For example, discrimination is unlikely to explain the paucity of articles by women in impeccably left-leaning publications that are staunchly committed to feminist precepts and gender equality.

References

Aronson, Joshua, Michael J. Lustina, Catherine Good, Kelli Keough, Claude M. Steele, and Joseph Brown. 1999. When White Men Can't Do Math: Necessary and Sufficient Factors in Stereotype Threat. *Journal of Experimental Social Psychology* 35: 29–46

Aronson, Joshua, and M. F. Salinas. 1997. Stereotype Threat, Attributional Ambiguity, and Latino Underperformance. Unpublished manuscript. Department of Psychology. University of Texas.

Baron-Cohen, Simon. 2003. *The Essential Difference: The Truth about the Male and Female Brain*. New York: Perseus Books Group.

Ben-Zeev, Talia, Steven Fein, and Michael Inzlicht. 2005. Arousal and Stereotype Threat. *Journal of Experimental Social Psychology* 41: 174–81.

Brown, Ryan R., and Robert A. Josephs. 1999. A Burden of Proof: Stereotype Relevance and Gender Differences in Math Performance. *Journal of Personality and Social Psychology* 76: 246–57.

Ceci, Stephen J., and Wendy M. Williams. 2006. Are We Moving Closer and Closer Apart? Shared Evidence Leads to Conflicting Views. In *Why Aren't More Women in Science?* ed. S. J. Ceci and W. M. Williams. Washington, D.C.: American Psychological Association.

Ceci, Stephen J., Wendy M. Williams, and Susan M. Barnett. 2009. Women's Underrepresentation in Science: Sociocultural and Biological Considerations. *Psychological Bulletin*, 135: 218–261

College Board. 2006. *Total Group Profile Report*. New York: College Board.

Croizet, Jean-Claude, and Theresa Clair. 1998. Extending the Concept of Stereotype Threat to Social Class: The Intellectual Underperformance of Students from Low Socioeconomic Backgrounds. *Personality and Social Psychology Bulletin* 24: 588–94.

Cullen, Michael J., Chiara M. Hardison, and Paul Sackett. 2004. Using SAT-Grade and Ability-Job Performance Relationships to Test Predictions Derived from Stereotype Threat Theory. *Journal of Applied Psychology* 89: 225.

Davies, Paul G., Steven J. Spencer, Diane M. Quinn, and Rebecca Gerhardstein. 2002. Consuming Images: How Television Commercials that Elicit Stereotype Threat Can Restrain Women Academically and Professionally. *Personality and Social Psychology Bulletin* 28: 1615–28.

Davies, Paul G., and Steven Spencer. 2004. The Gender-Gap Artifact. In *Gender Differences in Math*, ed. A. M. Gallagher and J. C. Kaufman. New York: Cambridge University Press.

Good, Catherine, Joshua Aronson, and Michael Inzlicht. 2003. Improving Adolescents' Standardized Test Performance: An Intervention to Reduce the Effects of Stereotype Threat. *Journal of Applied Developmental Psychology* 24: 645–62.

Halpern, Diane F., Camilla P. Benbow, David C. Geary, Ruben C. Gur, Janet Shibley Hyde, and Morton Ann Gernsbacher. 2007. The Science of Sex Differences in Science and Mathematics. *Psychological Science in the Public Interest* 8 (1): 1–51.

Hyde, Janet Shibley. 2006. Women in Science: Gender Similarities in Abilities and Sociocultural Forces. In *Why Aren't More Women in Science?* ed. S. J. Ceci and W. M. Williams. Washington, D.C.: American Psychological Association.

Inzlicht, Michael, and Talia Ben-Zeev. 2000. A Threatening Intellectual Environment: Why Females are Susceptible to Experiencing Problem-Solving Deficits in the Presence of Males. *Psychological Science* 11: 365–71.

———. 2003. Do High-Achieving Female Students Underperform in Private? The Implications of Threatening Environments on Intellectual Processing. *Journal of Educational Psychology* 95: 796–805.

Jencks, Christopher, and Melanie Phillips, eds. 1998. *The Black-White Test Score Gap.* Washington, D.C.: Brookings Institution Press.

Johns, Michael, Toni Schmader, and Andy Martens. 2005. Knowing Is Half the Battle: Teaching Stereotype Threat as a Means of Improving Women's Math Performance. *Psychological Science* 16: 175–79.

Journal of Blacks in Higher Education. 2006. A Large Black–White Scoring Gap Persists on the SAT. October 1. www.jbhe.com/features/53_SAT.html (accessed January 6, 2009).

Kelley, B. G. 2007. Why Teen Girls Outwrite Teen Boys. *Philadelphia Inquirer.* September 10, A11.

Levy, Becca. 1996. Improving Memory in Old Age through Implicit Self-Stereotyping. *Journal of Personality and Social Psychology* 71: 1092–1107.

Lubinski, David S., and Camilla P. Benbow. 2006. Sex Differences in Personal Attributes for the Development of Scientific Expertise. In *Why Aren't More Women in Science?* ed. S. J. Ceci and W. M. Williams. Washington, D.C.: American Psychological Association.

McGlone, Matthew S., and Joshua Aronson. 2006. Stereotype Threat, Identity Salience, and Spatial Reasoning. *Journal of Applied Developmental Psychology* 27: 486–93.

McIntyre, Rusty B., René M. Paulson, and Charles G. Lord. 2003. Alleviating Women's Mathematics Stereotype Threat through Salience in Group Achievement. *Journal of Experimental Social Psychology* 39: 83–90.

Morse, Stephen J. 2006. Brain Overclaim Syndrome and Criminal Responsibility: A Diagnostic Note. *Ohio State Journal of Criminal Law* 3: 397–412.

Murray, Charles. 2005. The Inequality Taboo. *Commentary.* September.

National Academy of Sciences, National Academy of Engineering, and Institute of Medicine of the National Academies. 2007. *Beyond Bias and Barriers: Fulfilling the Potential of Women in Academic Science and Engineering.* Washington, D.C.: National Academies Press.

New York Times. 2007. The 10 Best Books for 2007. December 9.

Nisbett, Richard R. 2009. Education Is All in the Mind. *New York Times.* February 8, News of the Week in Review op-ed page.

O'Brien, Laurie T., and Christian S. Crandall. 2003. Stereotype Threat and Arousal: Effects on Women's Math Performance. *Personality and Social Psychology Bulletin* 29: 782–89.

Park, Gregory, David Lubinski, and Camilla P. Benbow. 2008. Ability Differences among People Who Have Commensurate Degrees Matter for Scientific Creativity. *Psychological Science* 19: 957–61.

Perez-Pena, Richard. 2008. *Washington Post* Wins 6 Pulitzers; *Times* Reporting on Genetic Testing is Honored. *New York Times*. April 8, B7.

Quinn, Diane, and Steven Spencer. 2001. The Interference of Stereotype Threat with Women's Generation of Mathematical Problem-Solving Strategies. *Journal of Social Issues* 57: 55–71.

Sackett, Paul R., Chaitra M. Hardison, and Michael J. Cullen. 2004a. Comment—On the Value of Correcting Mischaracterizations of Stereotype Threat Research. *American Psychologist* 59 (1): 48–49.

———. 2004b. On Interpreting Stereotype Threat as Accounting for African American–White Differences on Cognitive Tests. *American Psychologist* 59 (1): 7–13.

———. 2005. On Interpreting Research on Stereotype Threat and Test Performance. *American Psychologist* 60 (3): 271–72.

Schmader, Toni. 2002. Gender Identification Moderates Stereotype Threat Effects on Women's Math Performance. *Journal of Experimental Social Psychology* 39: 194–201.

Spelke, Elizabeth, and Ariel Grace. 2006. Sex, Math, and Science. In *Why Aren't More Women in Science?* ed. S. J. Ceci and W. M. Williams. Washington, D.C.: American Psychological Association.

Spencer, Steven, Claude M. Steele, and Diane Quinn. 1999. Stereotype Threat and Women's Math Performance. *Journal of Experimental Social Psychology* 35: 4–28.

Steele, Claude M., and Joshua Aronson. 1995. Stereotype Threat and the Intellectual Test Performance of African Americans. *Journal of Personality and Social Psychology* 69: 797–811.

Wax, Amy L. 2004. The Threat in the Air. *Wall Street Journal*. April 13.

7

An Evolutionary Twist on Sex, Mathematics, and the Sciences

David C. Geary

For every woman in a department of mathematics, engineering, or physical sciences at an elite university in the United States, there are between seven and fourteen men.[1] In recent years, this sex difference has resulted in considerable media attention, as well as scientific reviews[2] and policy recommendations. The most prominent review is provided by the National Academy of Sciences' *Beyond Bias and Barriers: Fulfilling the Potential of Women in Academic Science and Engineering*, which concludes that "it is not a lack of talent, but unintentional biases and outmoded institutional structures that are hindering the access and advancement of women."[3] The report also suggests that sex differences in cognition—including spatial and mathematical abilities—and interests that are predictive of high levels of accomplishment in these fields are small, no longer relevant, or, if they do exist, not clearly related to biology or evolution.[4] It is ironic that, at the same time, several of the recommendations center on countering the disproportionate effects that childbirth and care—which are related to biology and evolution—have on the career trajectories of women.

In any case, I propose that in *Beyond Bias*, biological and potential evolutionary influences on sex differences were prematurely dismissed. This is not to say there are not talented women who can succeed at the highest levels in these fields, or that men have evolved to outperform women in these disciplines. Modern-day science and mathematics are evolutionarily novel, but this in and of itself does not preclude evolved but indirect influences on interest and acceleration in these fields.[5] I illustrate the gist in the following discussion.

Evolution

Over the past several decades, research in the biological sciences has led to substantial progress in our understanding of the evolution of sex differences[6] and of the mechanisms (such as sex hormones) that result in the here-and-now expression of these differences.[7] It is surprising that the insights gained from this considerable body of scientific research are not typically applied to the study of human sex differences. Biological influences, whether direct or indirect, by no means exclude social influences; in the same vein, social influences by no means exclude biological ones. The question is a matter of relative influence. At this time, we do not fully understand the degree to which various biological and social factors influence the outcomes described in the *Beyond Bias* report, in part because potential evolutionary and biological influences were not fully explored. Here I provide a primer on the mechanisms that drive the evolution of sex differences and those that influence the expression of these differences.

Sexual Selection. As described by Charles Darwin, the components of sexual selection and the primary mechanisms that drive the evolution of sex differences are *intrasexual competition*—competition with members of the same sex over mates—and *intersexual choice*—discriminative choice of mating partners.[8] Sex differences will evolve when the traits needed for successful intrasexual competition or that influence mate choice differ for males and females, and they typically do.[9] Figures 7-1 and 7-2 show examples of traits that have evolved as a result of male-male competition and female choice, respectively.

Sexual selection can also operate to create sex differences in cognitive abilities, as illustrated with comparisons of closely related species of vole (a kind of small rodent also known as field or meadow mice).[10] In the polygynous meadow vole (*M. pennsylvanicus*), males engage in scramble competition—that is, they compete by searching for and locating females that are dispersed throughout the habitat. Prairie and pine voles (*M. ochrogaster* and *M. Pinetorum*, respectively), in comparison, are monogamous, and males do not search for additional mates once paired. For the meadow vole, competition will favor males that court the most females, which is possible only through an expansion of the home range. Field studies of these voles have

FIGURE 7-1

THE MALE *ORYX LEUCORYX*

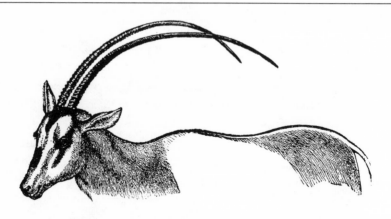

SOURCE: Darwin 1871, 251–52.
NOTES: During male–male competition, these males "kneel down, with their heads between their front legs, and in this attitude the horns stand nearly parallel and close to the ground. The combatants . . . endeavor to get the upturned points under each other's bodies; if one succeeds in doing this, he suddenly springs up, throwing up his head at the same time, and can thus wound or perhaps even transfix his antagonist."

confirmed the pattern shown in figure 7-3. Among meadow voles, males have home ranges that cover four to five times the area of the home ranges of females, but only during the breeding season,[11] as would be expected if range size is related to the reproductive strategy of the male (that is, its search for females). The home ranges of male and female prairie and pine voles overlap. A series of laboratory and field studies confirm the corollary prediction of a male advantage in spatial abilities for meadow voles, but no such sex difference for prairie and pine voles.[12]

Spritzer and others have confirmed that male meadow voles with good spatial abilities have larger home ranges and visit more females in their nests than do their less skilled peers.[13] Males with good spatial abilities typically sire more pups, but not always; female choice is also involved and is moderated by male aggressiveness and attention. On average, however, higher spatial abilities are associated with higher reproductive success—the predicted pattern, if this sex difference evolved as a result of male–male scramble competition.

FIGURE 7-2
FEMALE AND MALE HUMMINGBIRDS
(*SPATHURA UNDERWOODI*)

SOURCE: Darwin 1871, 77.
NOTES: The male's exaggerated tail feathers (right) are an indicator of physical health and the health of prospective offspring. Thus, females that choose these males as mates have more surviving offspring.

Sex Hormones. The expression of evolved sex differences is influenced by pre- and postnatal exposure to sex hormones, especially androgens such as testosterone.[14] Androgens influence sex differences in cognition and behavioral biases through early prenatal organization of associated brain areas or activation of these areas with postnatal exposure to androgens, or some combination. For male meadow voles, for instance, testosterone increases

FIGURE 7-3

MALE SCRAMBLE COMPETITION AND MALE MONOGAMY

Male Scramble Competition

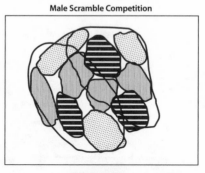

Male Monogamy

SOURCE: Author's illustration.

NOTES: With scramble competition (top panel), males compete by searching for females whose terri-tories (filled shapes) are dispersed. Territories of male meadow voles (bold, open shapes) encompass that of many females, and males may physically attack one another in regions in which territories overlap. The territories of male and female prairie and pine voles overlap (bottom panel).

significantly during the breeding season and spurs the increased activity lev-els needed to expand home ranges. For sexually selected behaviors that require an extended period of learning, testosterone or other hormones may act to increase engagement in these behaviors during development but might not in and of themselves result in adult-level competencies. In other words, hormones may result in a tendency to engage in one type of behavior or another, but behavioral or cognitive skill development often requires extended practice.

Within-Sex Variation. Sexual selection can exaggerate within-sex variation in the traits that influence competition and choice.[15] These traits are often condition-dependent—that is, their expression is heavily influenced by individual factors, such as health, and by social and ecological conditions during development and in adulthood. The resulting bias for extreme expression of these traits, as shown in figure 7-2, is maintained because their expression in fit individuals eliminates potential competitors that cannot express the traits to the same extreme. The basic point is that the sex that experiences more intense intrasexual competition or more intense vetting by members of the other sex will tend to show greater within-sex variation on many traits—variation that results, in part, from a sex difference in sensitivity to social or environmental conditions.

Human Evolution

Given this knowledge, it is surprising that evolutionary models of human sex differences are not often given full consideration, including in the *Beyond Bias* report. The development of evolutionary models of human sex differences is, of course, complex, but it is not simply the creation of "just so" stories. The use of patterns that emerge across species and human cultures, as well as studies of the proximate influences of sex hormones, provide methods to test the feasibility of these models.

Sexual Selection. A sex difference in physical size is a consistent indicator of an evolutionary history of sexual selection.[16] Larger males than females are nearly always associated with intense physical male–male competition and polygyny. The currently observed sex differences in physical size, upper-body musculature, patterns of physical development, and other traits strongly suggest the contribution of sexual selection to human evolution.[17]

Inferences about the nature of such sexual selection among humans are constrained by anthropological and population genetic studies. The details are nuanced and beyond the scope of this chapter,[18] but forms of male–male competition are relevant. In traditional societies, this competition includes coordinated group-level conflict for control of ecologically rich territories and for social and political influence.[19] Competition is often manifested in terms of low-level but frequent raiding, warfare, political manipulation, and result-ing high male mortality.[20] Within-group competition manifests physically and politically and results in the formation of dominance hierarchies.[21] Maintaining the groups' territorial borders, tribal warfare, and large-game hunting—all of which are almost exclusively male activities[22]—within these borders involves fluid movement in large and often novel ecologies. The result may be a male advantage in the complex three-dimensional spatial abilities that support large-scale navigation, as with meadow voles. The con-struction of weapons and other tools (also primarily a male activity) is also associated with hunting and warfare.

Cognitive Domains. The evolved function of behavior is to allow individ-uals to attempt to gain access to and control of the types of resources or avoid the types of threats that have tended to co-vary with survival or repro-ductive prospects during the species' evolutionary history.[23] At a broad level, most evolutionarily relevant resources or threats fall into three cate-gories: social, biological, and physical. The associated brain, cognitive, and other traits coalesce around the respective domains of folk psychology, folk biology, and folk physics.[24] Folk psychology is composed of the systems that enable people to negotiate social demands, and includes knowledge related to the self, dyadic relationships, and group-level interactions. Human folk biology supports the categorizing of plants and animals in the local ecology, as well as related knowledge that facilitates hunting and other activities involved in using these species as food and medicine.[25] Folk physics includes the systems that enable navigating in three-dimensional space and mentally representing this space (for example, a group's territory), and for using physical materials for tool-making.[26]

The knowledge and behavioral skills that compose folk systems emerge from an interaction of inherent constraints and experiences during devel-opment. Inherent constraints are most likely to evolve when the associated information, such as the basic shape of a human face, does not change from one generation to the next. Other folk-related information is more dynamic and results in some degree of novelty across generations and within life spans. The brain and cognitive systems that process this information can-not be highly constrained by biology, and hence are open to modification through experience.[27] In traditional societies it may take as long as twenty years to acquire the folk biological and physical knowledge needed for hunting or warfare.[28] Inherent bias in attention and interest evolve to initiate and guide the process of acquiring these competencies, but mastery requires a lot of experience.

Development and Interests. Children's social play and exploration of the environment and objects represent evolved tendencies to engage in certain activities, and they result in experiences that fine-tune folk competencies so they are well-suited to local conditions. Under different conditions, sex differences in developmental activities mirror those in patterns of intrasexual competition, intersexual choice, and parental investment.[29] These sex differences are influenced by pre- and postnatal exposure to sex hormones, but the developing competencies necessarily emerge from the interactions between early hormone-influenced biases in child-initiated activities and the specifics of the niches in which the children grow up. Socialization also influences the extent to which these differences are expressed—for instance, through suppression or encouragement of a desire for dominance and rough-and-tumble play.[30]

Play and other child-initiated activities related to the development of certain spatial abilities, as well as mechanical abilities associated with tool use, may provide a link to later emerging sex differences in some specific competencies related to mathematics and the sciences. For example, in both industrial and traditional societies, boys' play ranges are one and a half to three times the size of girls', and boys manipulate the ecology within this range more frequently and in more complex ways (such as by building forts) than do girls.[31] These activity differences result in a widening gap between boys and girls in the ability to visualize mentally and remember the geometric features of large-scale space—for instance, to remember and draw accurately the location of building A forty-five degrees northwest of building B. The sex difference in the size of the play range results from a combination of parental restrictions on girls' exploration away from home and child-initiated preferences that may be related to prenatal exposure to androgens, although the latter relationship is not fully understood.[32]

Most studies of infants and toddlers have failed to find sex differences in some object-related components of folk physics, such as discriminating mechanical from human motion.[33] There are, however, consistent sex differences in the play preferences of preschoolers and older children. Boys engage in more object play and girls in more family and parenting play.[34] Boys engage in more play with inanimate mechanical objects (such as toy cars) and play that involves building. In developing the preschool activities inventory—a survey in which adults report on the frequency with which children engage in various activities, such as doll play—Golombok and Rust obtained information on the play activities of 2,330 preschoolers from three nations.[35] The overall sex difference in early play activities was very large, and for individual activities it was largest for play with toy vehicles and toy weapons, which favored boys, and risk avoidance and pretending to be female characters, which favored girls. These findings were confirmed in a study of 3,990 preschool twins and their siblings.[36]

Boy-typical play preferences have also been found for girls with congenital adrenal hyperplasia (CAH), which results in excess prenatal exposure to androgens. For three- to eight-year-olds, Berenbaum and Hines reported that four out of five girls with CAH played with blocks, toy cars, and other "boys' toys" more often than their unaffected sisters or cousins. Three to four years later, the difference in preference for "boy play" was larger, such that nine out of ten girls with CAH engaged in more play with boys' toys than their unaffected relatives.[37] Within the normal range of early hormone exposure, Hines and colleagues found that higher maternal testosterone levels during pregnancy were associated with more masculine play for the preschool daughters of these mothers, but not for their sons.[38] For three-year-olds, Gredlein and Bjorklund found that engagement in a boy-typical form of object-oriented play was associated with skilled tool use during problem-solving for boys but not girls—boys benefited more from this type of play.[39] For eighteen-month-olds, Chen and Siegler found small to moderate advantages for boys for transfer of tool use from one setting to an analogous setting, in the consistency of tool use across settings, and in successful use of tools in problem-solving.[40]

Evolution, Mathematics, and the Sciences

Academic learning is possible, in part, because of the long developmental period among humans and the accompanying plasticity within folk domains, in combination with formal and informal instruction. We are only beginning to understand how this academic knowledge is built from the brain and cognitive systems that compose evolved folk domains, and the following is an early attempt to map such links.[41] I first illustrate several potential links between folk knowledge and academic learning in the sciences and mathematics, and then relate these to sex differences.

The Sciences and Mathematics. When asked about the forces acting on a thrown baseball, most people believe there are forces propelling it forward, something like an invisible engine, and propelling it downward. The downward force is gravity, but there is, in fact, no force propelling it forward, once the ball leaves the player's hand.[42] The concept of a forward force, called *impetus*, is similar to pre-Newtonian beliefs about motion prominent among intellectuals in the fourteenth to sixteenth centuries, and may reflect a feature of folk physics. Although adults can follow the trajectory of a thrown object, their *explicit* explanations of why the trajectory occurred as it did are often scientifically naïve.

Sir Isaac Newton's careful observation, use of the scientific method, and inductive and deductive reasoning resulted in a scientifically accurate understanding of large-scale motion and the operation of gravity. In his *Principia Mathematica*, Newton relied heavily on spatial and geometric representations to explain and prove his predictions about motion.[43] These spatial representations were likely dependent on the cognitive systems that evolved to support large-scale navigation. He built on this knowledge, however, and corrected naïve explanations through the use of explicit and exacting logic and a period of sustained effort and attention to this work.

The development of geometry—the study of space and shape—as a formal discipline may have also been influenced by early geometers' ability to explicitly represent the intuitive knowledge built into the systems that evolved to support navigation.[44] In the refinement and integration of the basic principles of geometry, Euclid formally and explicitly postulated that a straight line can be drawn from any point to any point; that is, he made explicit the intuitive understanding that the fastest way to get from one place to another is to go "as the crow flies." Using a few such basic postulates and definitions, Euclid then systematized existing knowledge to form the often complex and highly spatial components of classic geometry. As with the *Principia*, the creation of Euclid's *Thirteen Books of the Elements* must have required an exceptional ability to maintain attentional focus, along with well-developed spatial abilities and the ability to use logic to explicitly and precisely define spatial relationships.[45]

Linking Evolved Biases with Modern Competencies. My suggestion is that the historical emergence of mathematics and the sciences was built upon folk domains by individuals with high intelligence, spatial ability, creativity, ambition, and interest in folk physics.[46] High intelligence is associated with the ability to explicitly represent and manipulate folk and other information in working memory and to do so by means of formal logic. This ability, along with the development of the scientific method (and, thus, a means to test folk intuitions and correct naïve attributions) and the ability to transfer knowledge across generations (for instance, with books), was pivotal to this emergence. We are beginning to understand the brain and cognitive systems that allow people to explicitly represent information in working memory and to systematically manipulate this information in terms of formal rules and, through this, to build on the systems that appear to support folk domains.

As an example, brain imaging studies suggest that areas of the parietal cortex comprise one link between evolved spatial abilities and some aspects of mathematical learning.[47] The parietal cortex also appears to be part of the brain systems that support the mental simulation of how objects can be manipulated when used as tools.[48] Although a functional relationship can only be guessed at, it is of interest that areas of the parietal cortex typically associated with spatial imagery and other areas of folk physics were unusually large in Albert Einstein's brain.[49]

My suggestion is that being at the extreme of the folk-physical systems that support the use of spatial imagery contributes—likely in combination with areas of the prefrontal cortex—to the ability to explicitly represent physical and complex three-dimensional, spatially based quantitative information in working memory, and thus may contribute to the ease of learning spatial and mechanical aspects of mathematics and the sciences.[50] Preliminary evidence suggests that enhanced folk-physical systems may contribute to eminent contributions in these fields.[51]

Sex Differences

Having considered the evolution of sex differences generally, I next provide some hypotheses linking evolved sex differences to those that emerge in the sciences and mathematics.

Brain and Cognition. The potential relationship between the brain and cognitive systems for navigation and geometry, as well as the use of spatial representations for solving some types of mathematical problems, means that any sex differences in activities that enhance these navigation-related spatial abilities may incidentally contribute to sex differences in these areas of mathematics. If the evolved activity preferences of boys result in elaboration of these spatial systems during development—and they appear to[52]—then boys may have a head start in setting the cognitive foundation for learning in some mathematical and scientific areas. This does not mean formal training with use of spatial representations cannot close the gap between males and females, but in the absence of such training, boys will engage in more informal, child-initiated spatial activities than girls.

The relationship between activity in areas of the parietal cortex and certain spatial abilities may provide a place to begin the systematic testing of this type of hypothesis. Based on a structural magnetic resonance imaging (MRI) study, Goldstein and colleagues found many of these brain regions 20–25 percent larger in men than in women, with a high density of sex hormone receptors during prenatal development.[53] One implication is that the cognitive and behavioral functions supported by these regions were under stronger selection pressures for men than for women during human evolution, and that their organization may be influenced by prenatal exposure to sex hormones. While a relationship has not been established between the parietal cortex and sex differences in geometry and mathematical reasoning, it is an area of potential future study.

Interests and Ambition. As described in "Development and Interests," above, boys are more likely than girls to show an interest in and engage in activities that might enhance folk-physical competencies. A corresponding hypothesis is that these interest biases reflect, in part, more frequent tool construction by males than females during human evolution. They are thought to be an aspect of the suite of brain and cognitive systems and developmental activities that allowed males to explore and learn how to construct and use tools from materials in the local ecology. If these sex differences contribute to the sex differences in interest in some areas of the physical sciences and engineering, then within- and between-sex differences in these interests should follow from a pattern of interest and engagement in play with mechanical objects during development, and from prenatal exposure to male hormones.[54]

In addition to area-specific interests, an evolutionary history of more intense intrasexual competition and intersexual vetting in males than females will manifest in modern societies as men being more focused than women on cultural status in the domains in which they compete.[55] In societies with economic specialization, a corollary is that men will define success in terms of areas in which they have a competitive advantage over other men and may accordingly be narrow in their interests and the focus of where they compete. From this perspective, men should also be more willing than women to trade off time with friends and family to gain status in these areas—a prediction that has been confirmed with Lubinski and Benbow's study of mathematically gifted people.[56] The willingness to make these trade-offs in combination with an interest in mathematics and science are two essential components of exceptional accomplishment.

Within-Sex Variation. Based on features of male–male competition, greater within-sex variation should follow from three-dimensional spatial cognition in boys and men. If the development of spatial cognition is a condition-dependent trait in males, then boys should benefit more than girls in environments that enable the play-based exploration of large-scale space, and to suffer more than girls in environments that restrict these activities. Levine and colleagues assessed a sample of 547 children from high-, middle-, and low-income backgrounds across second and third grades on two spatial tasks and a syntax comprehension test.[57] While there were no sex differences on the syntax test, boys had an advantage on both spatial tasks—but only among children from high- and middle-income families. There were no sex differences for the low-income children. In other words, low family income was associated with lower scores for both boys and girls on all three tests, but in relationship to same-sex peers, the spatial competencies of boys were more strongly affected by poverty than those of girls. The features of these environments that influenced spatial development are not known, and thus this is not a strong test of the greater sensitivity of boys to early experiences. The pattern is nevertheless consistent with evolutionary predictions regarding the influence of environmental circumstance on within-sex variation in traits that are potentially related to sexual selection, and it illustrates the richness of this approach for understanding potential environmental influences on the expression of evolved sex differences.

Conclusion

Evidence that sexual selection is responsible for sex differences across many species is now overwhelming.[58] The dynamics result in the exaggeration of traits that facilitate intrasexual competition or influence mate choice, and in so doing they create between-sex differences and often increase within-sex variation as well. The here-and-now expression of sexually selected traits results from a combination of pre- and postnatal exposure to sex hormones, genetic sex, and ecological and social contexts.[59] Some of these traits are condition-dependent—that is, they have evolved to reflect how individuals of one sex or the other are coping with ecological stressors (for example, parasites) and social stressors (for example, competition for status). Thus, expression of the trait is highly dependent on experience and context.

Human sex differences follow from the extent to which these components of sexual selection have differed for males and females during human evolution. For instance, male–male competition includes group-level warfare, tool construction, and use of projectile weapons, among other things, that are not components of female–female competition.[60] Based on these differences, males have an advantage in use of geometric-based navigational strategies for movement in novel ranges and for the forms of mechanics-related cognition that support tool construction. Any such sex differences emerge slowly during childhood and are influenced by pre- and postnatal exposure to sex hormones, as well as by the influence of those hormones on child-initiated activities that promote learning in these areas.

Learning the knowledge bases, technical skills, and conceptual models that compose mathematics and the sciences is necessarily related to schooling, and not directly to cognitive evolution. Nevertheless, evolved cognitive abilities and attendant biases in interests provide the foundation for academic learning[61] and can add to our understanding of how people learn and understand mathematics and the sciences, and why they might pursue careers in these fields. Sex differences in folk domains follow from the extent to which these domains are related to parenting or to patterns of intrasexual competition or mate choice. During human evolution, male–male competition was almost certainly dependent on spatial abilities, and in some mechanical domains related to tool construction. Thus, we can expect average sex differences to favor boys and men, as well as the presence of more boys and men at the extremes of these abilities and associated interests (for instance, in mechanical objects). We can also expect sex differences in mastering areas of mathematics and the sciences in which these forms of folk cognition might influence learning.

This perspective helps to frame the relationship between spatial cognition and mathematical reasoning and the male advantage in both of these areas.[62] Any such relationship does not lead to a blanket prediction of a male advantage in all areas of mathematics, but rather provides specific and testable hypotheses about where the sexes should be similar and where they should be different; for instance, we can predict a male advantage in the ability to visualize mathematical relationships in three dimensions, as well as no sex difference in the ability to learn geometric theorems. Another implication is that formal training in use of spatial strategies for solving mathematics problems may be more important for girls than for boys, and that such training may help to close the gap. My point is that an evolutionary and biological approach has much to add to our understanding of human sex differences, including those that contribute to the overrepresentation of men in mathematics, engineering, and the physical sciences.

In this light, the *Beyond Bias* report would have better served the nation with a full and thorough consideration of all available and relevant scientific knowledge of potential influences on the observed sex differences in the physical sciences, mathematics, and related fields. Without such consideration, policy initiatives resulting from the *Beyond Bias* report are at risk for failure, at best, and harm to the scientific infrastructure of the United States, at worst. The stakes are too high to attempt to institute policy and institutional change without a *full* understanding of sex differences in the development of scientific and mathematical talent and in the long-term progression of men and women in these fields. The *Beyond Bias* report does not provide this foundation.

Notes

1. Halpern et al. 2007.
2. Ibid.; Ceci and Williams 2006; National Academy of Sciences et al. 2007; Spelke 2005.
3. National Academy of Sciences et al. 2007, 1.
4. Lubinski and Benbow 2006; Webb et al. 2007.
5. Geary 1996 and 2007.
6. Andersson 1994.
7. Adkins-Regan 2005.
8. Darwin 1871.
9. Andersson 1994.
10. Gaulin and Fitzgerald 1986 and 1989; Spritzer et al. 2005.
11. Gaulin and Fitzgerald 1986.
12. Gaulin and Fitzgerald 1986 and 1989.
13. Spritzer et al. 2005.
14. Morris et al. 2004.
15. Pomiankowski and Møller 1995.
16. Andersson 1994.
17. Tanner 1990.
18. Geary 1998.
19. Chagnon 1988.
20. Keeley 1996.
21. For more on dominance hierarchies, see chapter 1, above.
22. Murdock 1981.
23. Geary 2005.
24. Hirschfeld and Gelman 1994.
25. Atran 1998.
26. Pinker 1997; Shepard 1994. For more on representation of objects, see chapter 2, above.
27. Geary 2005.
28. Kaplan et al. 2000.
29. Geary 1998.
30. Low 1989.
31. Matthews 1992.
32. Resnick et al. 1986; Hines et al. 2003; Whiting and Edwards 1988.
33. Spelke 2005.
34. Cohen-Bendahan et al. 2005; Maccoby and Jacklin 1974.
35. Golombok and Rust 1993.
36. Iervolino et al. 2005.
37. Berenbaum and Hines 1992.
38. Hines et al. 2002.

39. Gredlein and Bjorklund 2005.
40. Chen and Siegler 2000.
41. Geary 2007.
42. Clement 1982.
43. Newton [1687] 1995.
44. Geary 1995.
45. Euclid [c. 300 BC] 1956.
46. Geary 2007; Murray 2003.
47. Dehaene et al. 1999.
48. Johnson-Frey 2003.
49. Witelson et al. 1999.
50. Baron-Cohen 2003.
51. Baron-Cohen et al. 1999.
52. Matthews 1992.
53. Goldstein et al. 2001.
54. Baron-Cohen 2003.
55. Geary 1998.
56. Lubinski and Benbow 2006.
57. Levine et al. 2005.
58. Andersson 1994; Darwin 1871.
59. Adkins-Regan 2005.
60. Geary 1998.
61. Geary 2007.
62. Geary 1996.

References

Adkins-Regan, Elizabeth. 2005. *Hormones and Animal Social Behavior.* Princeton, N.J.: Princeton University Press.

Andersson, Malte. 1994. *Sexual Selection.* Princeton, N.J.: Princeton University Press.

Atran, Scott. 1998. Folk Biology and the Anthropology of Science: Cognitive Universals and Cultural Particulars. *Behavioral and Brain Sciences* 21: 547–609.

Baron-Cohen, Simon. 2003. *The Essential Difference: The Truth about the Male and Female Brain.* New York: Perseus Books Group.

Baron-Cohen, Simon, Sally Wheelwright, Valerie Stone, and Melissa Rutherford. 1999. A Mathematician, a Physicist and a Computer Scientist with Asperger Syndrome: Performance on Folk Psychology and Folk Physics Tests. *Neurocase* 5: 475–83.

Berenbaum, Sheri A., and Melissa Hines. 1992. Early Androgens Are Related to Childhood Sex-Typed Toy Preferences. *Psychological Science* 3: 203–6.

Ceci, Stephen J., and Wendy W. Williams, eds. 2006. *Why Aren't More Women in Science? Top Researchers Debate the Evidence.* Washington, D.C.: American Psychological Association.

Chagnon, Napoleon A. 1988. Life Histories, Blood Revenge, and Warfare in a Tribal Population. *Science* 239 (4843): 985–92.

Chen, Zhe, and Robert S. Siegler. 2000. Across the *Great Divide: Bridging the Gap Between Understanding Toddlers' and Older Children's Thinking.* Chicago, Ill.: Wiley-Blackwell.

Clement, John. 1982. Students' Preconceptions in Introductory Mechanics. *American Journal of Physics* 50: 66–71.

Cohen-Bendahan, Celina C. C., Cornelieke van de Beek, and Sheri A. Berenbaum. 2005. Prenatal Sex Hormone Effects on Child and Adult Sex-Typed Behavior: Methods and Findings. *Neuroscience and Biobehavioral Reviews* 29: 353–84.

Darwin, Charles. 1871. *The Descent of Man, and Selection in Relation to Sex.* London: John Murray.

Dehaene, Stanislas, Elizabeth Spelke, Philippe Pinel, Ruxandra Stanescu, and Sanna Tsivkin. 1999. Sources of Mathematical Thinking: Behavioral and Brain-Imaging Evidence. *Science* 284 (5416): 970–74.

Euclid. [c. 300 BC] 1956. *The Thirteen Books of the Elements.* Vol. 1. Trans. T. L. Heath. New York: Dover.

Gaulin, Steven J. C., and Randall W. Fitzgerald. 1986. Sex Differences in Spatial Ability: An Evolutionary Hypothesis and Test. *American Naturalist* 127: 74–88.

———. 1989. Sexual Selection for Spatial-Learning Ability. *Animal Behaviour* 37: 322–31.

Geary, David C. 1995. Reflections of Evolution and Culture in Children's Cognition: Implications for Mathematical Development and Instruction. *American Psychologist* 50: 24–37.

———. 1996. Sexual Selection and Sex Differences in Mathematical Abilities. *Behavioral and Brain Science* 19: 229–84.

———. 1998. *Male, Female: The Evolution of Human Sex Differences*. Washington, D.C.: American Psychological Association.

———. 2005. *The Origin of Mind: Evolution of Brain, Cognition, and General Intelligence*. Washington, D.C.: American Psychological Association.

———. 2007. Educating the Evolved Mind: Conceptual Foundations for an Evolutionary Educational Psychology. In *Educating the Evolved Mind Vol. 2, Psychological Perspectives on Contemporary Educational Issues*, ed. J. S. Carlson and J. R. Levin, 1–99. Greenwich, Conn.: Information Age.

Goldstein, Jill M., Larry J. Seidman, Nicholas J. Horton, Nikos Makris, David N. Kennedy, Verne S. Caviness Jr., et al. 2001. Normal Sexual Dimorphism of the Adult Human Brain Assessed by In Vivo Magnetic Resonance Imaging. *Cerebral Cortex* 11: 490–97.

Golombok, Susan, and John Rust. 1993. The Pre-School Activities Inventory: A Standardized Assessment of Gender Role in Children. *Psychological Assessment* 5: 131–36.

Gredlein, Jeffrey M., and David F. Bjorklund. 2005. Sex Differences in Young Children's Use of Tools in a Problem-Solving Task. *Human Nature* 16: 211–32.

Halpern, Diane F., Camilla P. Benbow, David C. Geary, Ruben C. Gur, Janet Shibley Hyde, and Morton Ann Gernsbacher. 2007. The Science of Sex Differences in Science and Mathematics. *Psychological Science in the Public Interest* 8 (1): 1–52.

Hines, M., B. A. Fane, V. L. Pasterski, G. A. Mathews, G. S. Conway, and C. Brook. 2003. Spatial Abilities Following Prenatal Androgen Abnormality: Targeting and Mental Rotations Performance in Individuals with Congenital Adrenal Hyperplasia. *Psychoneuroendocrinology* 28: 1010–26.

Hines, M., Susan Golombok, John Rust, Katie J. Johnston, Jean Golding, and the Avon Longitudinal Study of Parents and Children Team. 2002. Testosterone during Pregnancy and Gender Role Behavior in Preschool Children: A Longitudinal Population Study. *Child Development* 73 (6): 1678–87.

Hirschfeld, Lawrence A., and Susan A. Gelman, eds. 1994. *Mapping the Mind: Domain Specificity in Cognition and Culture*. New York: Cambridge University Press.

Iervolino, Alessandra C., Melissa Hines, Susan E. Golombok, John Rust, and Robert Plomin. 2005. Genetic and Environmental Influences on Sex-Typed Behavior during the Preschool Years. *Child Development* 76: 826–40.

Johnson-Frey, Scott H. 2003. What's So Special About Human Tool Use? *Neuron* 39:201–4.

Kaplan, Hillard, Kim Hill, Jane Lancaster, and A. Magdalena Hurtado. 2000. A Theory of Human Life History Evolution: Diet, Intelligence, and Longevity. *Evolutionary Anthropology* 9: 156–85.

Keeley, Lawrence H. 1996. *War before Civilization: The Myth of the Peaceful Savage*. New York: Oxford University Press.

Levine, Susan C., Marina Vasilyeva, Stella F. Lourenco, Nora Newcombe, and Janellen Huttenlocher. 2005. Socioeconomic Status Modifies the Sex Differences in Spatial Skill. *Psychological Science* 16: 841–45.

Low, B. S. 1989. Cross-Cultural Patterns in the Training of Children: An Evolutionary Perspective. *Journal of Comparative Psychology* 103: 311–19.

Lubinski, David, and Camilla P. Benbow. 2006. Study of Mathematically Precocious Youth after 35 Years: Uncovering Antecedents for the Development of Math-Science Expertise. *Perspectives on Psychological Science* 1: 316–45.

Maccoby, Eleanor E., and Carol N. Jacklin. 1974. *The Psychology of Sex Differences.* Stanford, Calif.: Stanford University Press.

Matthews, M. H. 1992. *Making Sense of Place: Children's Understanding of Large-Scale Environments.* Savage, Md.: Barnes and Noble Books.

Morris, John A., Cynthia L. Jordan, and S. Marc Breedlove. 2004. Sexual Differentiation of the Vertebrate Nervous System. *Nature Neuroscience* 7: 1034–39.

Murdock, George Peter. 1981. *Atlas of World Cultures.* Pittsburgh, Pa.: University of Pittsburgh Press.

Murray, Charles. 2003. *Human Accomplishment: The Pursuit of Excellence in the Arts and Sciences, 800 B.C. to 1950.* New York: HarperCollins.

National Academy of Sciences, National Academy of Engineering, and Institute of Medicine of the National Academies. 2007. *Beyond Bias and Barriers: Fulfilling the Potential of Women in Academic Science and Engineering.* Washington, D.C.: National Academies Press.

Newton, Isaac. [1687] 1995. *The Principia Mathematica.* Trans. A. Motte. Amherst, N.Y.: Prometheus Books.

Pinker, Steven. 1997. *How the Mind Works.* New York: W. W. Norton.

Pomiankowski, A., and A. P. Møller. 1995. A Resolution of the Lek Paradox. Proceedings of the Royal Society of London B260: 21–29.

Resnick, Susan M., Sheri A. Berenbaum, I. I. Gottesman, and T. J. Bouchard Jr. 1986. Early Hormonal Influences on Cognitive Functioning in Congenital Adrenal Hyperplasia. *Developmental Psychology* 22: 191–98.

Shepard, Roger N. 1994. Perceptual–Cognitive Universals as Reflections of the World. *Psychonomic Bulletin and Review* 1: 2–28.

Spelke, Elizabeth S. 2005. Sex Differences in Intrinsic Aptitude for Mathematics and Science: A Critical Review. *American Psychologist* 60: 950–58.

Spritzer, Mark D., Nancy G. Solomon, and Douglas B. Meikle. 2005. Influence of Scramble Competition for Mates Upon the Spatial Ability of Male Meadow Voles. *Animal Behaviour* 69: 375–86.

Tanner, J. M. 1990. *Fetus into Man: Physical Growth from Conception to Maturity.* Cambridge, Mass.: Harvard University Press.

Webb, Rose Mary, David Lubinski, and Camilla P. Benbow. 2007. Spatial Ability: A Neglected Dimension in Talent Searches for Intellectually Precocious Youth. *Journal of Educational Psychology* 99 (2): 397–420.

Whiting, Beatrice B., and Carolyn Edwards. 1988. *Children of Different Worlds: The Formation of Social Behavior.* Cambridge, Mass.: Harvard University Press.

Witelson, Sandra F., Debra L. Kigar, and Thomas Harvey. 1999. The Exceptional Brain of Albert Einstein. *Lancet* 353: 2149–53.

8

Cognition and the Brain: Sex Matters

Richard J. Haier

Research to determine the nature of cognition has a long and honorable history. Scientific programs to understand the basic processes of cognition—attention, learning, and memory—generally are well-supported and without much public controversy. Cognition is based in the brain, and many researchers around the world are engaged in identifying precisely how the brain manages it. A special urgency is associated with understanding how cognition is disrupted when the brain is attacked by Alzheimer's disease, schizophrenia, stress, aging, and a host of other diseases and insults. Often, researchers interested in dissecting normal cognition find funding only by invoking the promise of their research to elucidate some aspect of abnormal cognition and disease.

It is fair to characterize most research on cognition as normative—that is, its goal is to establish the general principles that underlie basic cognitive processes like learning and memory. Normative research generally focuses on what is common among subjects, whether rats, mice, or humans. If, for example, a drug given to a group of laboratory rats disrupts their subsequent average performance, compared to placebo and control conditions, on a problem-solving task previously memorized, this is evidence that the drug affects the aspect of memory in question.

Researchers who focus on individual differences in cognition, on the other hand, try to address such questions as whether each rat receiving the drug shows the same magnitude of effect on memory. What does it mean if some rats have a very strong effect and others show a weak or zero effect, especially if all the rats are of the same genetic strain and were raised in similar environments? Here is where controversy begins in this exceedingly complex and important work. After a hundred years of laboratory experiments on learning and memory, we still question whether experimental findings can be generalized to answer why some children learn faster or memorize more than others, or why some people reason better than others.

Research on individual differences in cognition has a controversial history. This is to be expected somewhat because of the focus on how we differ from each other, rather than on what we have in common. In fact, research on intelligence has one of the most controversial histories in science, even though intelligence, as commonly understood, refers to little more than individual differences in basic cognitive processes, especially learning and memory. Along with the usual and necessary scientific skepticism and debate about various findings have come emotional and political criticisms, not unlike those targeted against stem cell research. The combination of the study of genetics and/or biology (the basis for modern neuroscience) with research investigations examining the sources of variance for individual differences can be incendiary. In 2005, for example, the president of Harvard University, Lawrence Summers, suggested that, in addition to discrimination and other sociocultural factors, there might also be biological reasons for fewer women than men having careers in science and math. The response to his remarks, especially in academia, was a vociferous and acrimonious outcry and the sudden, compelling desire of the president of Harvard to spend more time with his family. As late as September of 2007, Dr. Summers was abruptly disinvited to speak at a dinner of the University of California Regents because of some faculty complaints concerning his remarks two years previously.[1]

Uncertainty in the Field

The furor over Summers's remarks also resulted in renewed interest and debate among scientists about what the data on sex differences and cognition actually show and what they mean. One group of prominent researchers recently reviewed the data from several perspectives.[2] They reached a consensus that there are some sex differences in cognitive ability, and that many factors, both social and biological, contribute to career choices in science and math professions. Their summary of the data and the issues, including definitions of key concepts and terms, is succinct and demonstrates why simple conclusions are not yet possible. Another comprehensive and balanced presentation of the data and their interpretation by many experts on this issue can be found in a collection of papers published in 2006 by the American Psychological Association, entitled, *Why Aren't More Women in Science?*[3] In its concluding chapter, the editors summarize several key issues, focusing on areas where the same data are interpreted differently by researchers. Among them are the following four:

- *The right tail of the distribution.* On average, males and females are essentially equivalent on most measures of cognitive performance, although some differences in specific abilities are consistently found. An important question is whether there are more males at the very highest end of the distribution of science and math ability. This argument centers not on superior ability in the top 5 percent or even the top 1 percent, where sex differences are not large and may have no practical implication for professional success. Rather, the question is about extraordinary ability in the top one-tenth of 1 percent. Some of the best evidence about this rarified group (that is, about 1 of 10,000 individuals) comes from longitudinal studies of mathematically precocious youth begun at Johns Hopkins University more than thirty years ago. Based largely on the results of the SAT-M test, this research shows more boys than girls in the extraordinary range. There is no definitive explanation for why more boys than girls score in the top one-tenth of 1 percent, but some think that the use of the SAT-M for this categorization may not provide an accurate reflection of underlying aptitude, as opposed to manifest performance. Others argue that performance tests always assess aptitude, at least to some degree. Experts also disagree as to whether the magnitude of the performance difference (three to one in recent data, down from thirteen to one in the early 1970s data), suggests a biological basis.[4]

- *Real-world demands.* Gender-related real-world demands are a subject of much interest with respect to explaining why there are fewer women in certain fields. Whereas experts generally agree that the most successful professional achievement requires the dedication to it of over sixty hours a week, data indicate that women work fewer hours in their professions because of competing cultural roles demanding more of them for childrearing and family responsibilities. Some data also suggest that women who work fewer hours end up being more satisfied. As is often the case, especially when using survey data, conflicting interpretations are frequent.[5]

- *Differing preferences.* Some data indicate that differing career interests may account for fewer women than men choosing certain professions, suggesting that, on average, women prefer people-oriented professions (such as medicine and law), while men prefer object-oriented ones (such as engineering and physics). Interpretations of these data also differ. At best, the factors shaping any such differences in career interests are not well established empirically.[6]

- *Interaction of biology and environment.* A small biological advantage for some attribute may lead a person to seek out environments in which the advantage comes to flourish. Disentangling such effects is difficult. They may contribute to fewer women being in certain professions, but this is not determined.[7]

These examples underscore the uncertainty among researchers regarding the existence and importance of sex differences in cognition as they may relate to professional choices and success. Ceci and Williams wonder how this research can even proceed when some interpretations of data, especially those suggesting some biological basis for some sex differences, are offensive to many people. They conclude that "all legitimate views need to be aired openly for science to flourish and for policies to be well informed by scientific findings."[8]

Certainty at the NAS

A committee of the National Academy of Sciences (NAS) has taken a different position. Their report on women in science, *Beyond Bias and Barriers: Fulfilling the Potential of Women in Academic Science and Engineering*,[9] includes specific recommendations to address the paucity of women in the physical sciences and engineering. While summarizing essentially the same research literature covered by Ceci and Williams's book and the review by Halpern and colleagues, the NAS report finds far less nuance and uncertainty in the data. It asserts that sex differences in cognition, if they exist at all, do not have any bearing on why fewer women are in physical and engineering professions, and reasons that if there are no sex differences in cognition, there is no need to argue over their cause. The report endorses support for additional research on social and cultural factors to address the gender imbalance in certain professions; there is no endorsement for more biological or neuroscience research.

Here is the committee's logic, as expressed in the opening summary of chapter 2 of the report, entitled, "Learning and Performance":

> 1. Most research studies of sex differences in cognition find more overlap than differences and any differences tend to be small. 2. When average differences are found, the differences are on measures designed to predict high school and college success; and since there is no longer any academic performance difference between male and female students, there is no gender gap to explain. 3. The observation that there are more males than females in the very top tier of performance on tests of mathematical reasoning is based on tests (e.g. the SAT-M) designed to predict high school and college success, not success in science careers. 4. *Thus,* we can not look to cognitive sex differences to explain the differential success of men and women scientists and engineers.[10]

The data interpretations given to justify this logic are not compelling. Virtually all researchers in this field agree with point 1, but questions remain open with respect to aspects of spatial ability and their relationship to mathematical reasoning, a key component of science and engineering. Point 2 also is widely accepted for most cognitive abilities, but the same questions about spatial ability remain, especially about the top one-tenth of 1 percent of students. Point 3 is based on a frequently used ploy to blame the test when results are controversial; more importantly, the SAT-M specifically has been shown to predict professional success.[11] As to point 4, despite the NAS committee's assertion, research questions about sex differences in cognition that may have bearing on the imbalance of women in certain professions remain unresolved.[12]

The NAS report also all but eliminates the possibility that any biological factors may be relevant to understanding the complexities of professional success. The report does acknowledge anatomical and functional differences between the brains of males and females, but asserts that "studies of brain structure and function . . . have not revealed biological differences between men and women in performing science and mathematics that can account for the lower representation of women in these fields."[13] The wording of this assertion is overly precise—what, exactly, would such brain studies look like? Have any studies, positive or negative, been directed at this exact wording of the problem?

Some Relevant Brain Data

It would be odd to assume that cognition had nothing to do with the biology of the brain, and it would be even odder to assume that individual differences in brain biology were not related to individual differences in cognition. Tools to study the brain directly have only become available to cognition researchers relatively recently. Neuroimaging techniques are especially powerful, and many researchers are now engaged in determining how brain structure and function are related to cognition. Thousands of these studies are published; most are directed at normative processes and do not investigate individual differences, and most do not address any male/female differences.

There are exceptions in which male/female differences are specifically investigated; some of these studies are even cited in the NAS report.[14] Haier and Benbow first addressed the question in 1995, when we used functional brain imaging with positron emission tomography (PET) to examine the brain areas used for mathematical reasoning.[15] We studied twenty-two male and twenty-two female college students. Half of each group was selected for having had college entrance SAT-M scores over 700, and the other half for having had average scores. All subjects were scanned while they performed a new SAT-M test. Brain activity in specific parts of the temporal lobe was correlated to SAT-M score in the men, but not in the women. This was a clear difference. Note that even though the key research design element in this study was selecting male and female subjects matched for mathematical reasoning ability, the data still showed a strong sex difference in the brain. This methodology took the investigation of biological factors and sex differences in cognition in a new direction by investigating whether different brain architectures or designs produced equal cognitive performance.

Recently, we reported a similar finding using structural magnetic resonance imaging (MRI) to assess gray and white matter throughout the brain and correlating the amount of tissue with IQ scores from the Wechsler Adult Intelligence Scale (WAIS). We found significant correlations for several brain areas,[16] but the areas differed for men and for women matched on IQ scores and controlling for brain size.[17] In men, more gray matter was related to IQ; in women, white matter was more important. These results suggested that the brain might have more than one organization designed for intelligence.[18] Interestingly, one of the areas which appeared to be more important for intelligence in men than in women was in the parietal lobe, an area important for visual-spatial ability; it is an area where Albert Einstein's brain, at autopsy, showed greater volume than control brains.[19]

Based on a review of thirty-seven neuroimaging studies of intelligence published since 1988, a specific network of brain areas has been proposed as a neural basis for intelligence,[20] but there are not yet enough studies to tell us if the same network details apply to both males and females. The study of brain structure and function in individuals selected for extraordinary ability in a specific cognitive domain may be particularly informative. More than one brain design or architecture may be identified, and each may be related to high cognitive performance. Research to determine the frequency of these hypothesized brain designs in males and females, matched for extraordinary ability, may be important in understanding the depth of the career disparity problem. Even matching males and females for average performance at the fiftieth percentile may yield different brain architectures. Such studies done in children surely would elucidate important developmental factors.[21] Even a passing familiarity with neuroimaging studies like these demonstrates that there is so much we do not know and so much yet to discover about brain biology and sex differences, and perhaps even career choices.

Offense and Myth

The NAS committee's discussion of biology concludes, "Because men and women do not differ in their average abilities and because they have now achieved equal academic success in science through the college level, there is no sex performance difference for the biological studies and theories to explain."[22] This logic is not compelling for such a definitive closure. What about the observed fact of fewer women in science and engineering professions after college? Could brain differences between men and women emerge in later life when some brain-related genes turn on or off? Are there any biological factors at all which could have even a little to do with the disparity? Given the sophistication of neuroimaging technology and other neuroscience techniques, would it not be appropriate for the NAS committee to call for more research rather than for none? Whatever social and cultural factors help explain gender differences in science professions, and there surely are important ones, why must the possibility that any biological factors contribute even a little to the disparity be excluded so definitively? It is likely that, as Ceci and Williams acknowledged, some people are offended by the possibility that biological factors, genetic or not, may be involved. Such offense is often based on misunderstanding or myth about what it means if something is biological.

There is no need for such offense. Suppose a human attribute "X" is the result of biological processes. This does not mean that X is destined to be fixed in stone, even if there is a genetic component to the biological process (not every biological phenomenon is genetic, but anything genetic always works through biology, whether through interaction with the environment or not). Moreover, the possibility that X may actually change is not the basis for a logical argument that X cannot be biological or even genetic. Biology can be changed, as is apparent every time you see your doctor and ask for broken biology to be fixed. If X has a genetic component, it may not be expressed equally in all populations, and the frequency of genes may differ; that is why there are more blue-eyed people in some countries than in others. Inferences that X confers inferior or superior status on a person or a group are judgmental and inappropriate, whether X has a biological basis or not. When it comes to X, interactions between brain and environment are two-way streets (this used to be called the nature/nurture debate). More knowledge of how the interactions work in both directions is always better than less.

It is surprising and unfortunate that myths and misunderstandings surrounding such issues are well represented in chapter 2 of the NAS report.[23] Efforts of neuroscience to understand the biology of how the brain works, both in general and in individuals, face serious and important questions whose resolution requires unprejudiced scientific approaches to understanding sex differences. We may find, for instance, that men and women of equal cognitive ability use different brain mechanisms to achieve the same performance. If so, we have learned something potentially important for cognitive rehabilitation following brain injury, and for minimizing the effects of normal aging or brain disease. We may also learn that one brain mechanism is more common than another in either sex. This could have a bearing on different rates of career choices and success. Diane Halpern, a leading researcher on cognitive differences between men and women, is quoted in the NAS report:

> Some researchers object to the study of sex differences because they fear that it promotes false stereotypes and prejudice. There is nothing inherently sexist in a list of cognitive sex differences; prejudice is not intrinsic in data, but can be seen in the way people misuse data to promote a particular viewpoint or agenda. Prejudice also exists in the absence of data. Research is the only way to separate myth from empirically supported findings.[24]

It is disappointing that the NAS report apparently ignores this sentiment when it comes to the exciting potential of neuroscience investigations for understanding more about sex differences.

Conclusion

We just do not have simple answers about the observed disparity in the presence of men and women in science and engineering professions because there are compelling data that suggest several interacting factors, some cultural and social, others biological. Objective readers will understand that no one study or perspective will provide a satisfactory understanding of the complexities of career choice and professional success. Many may conclude that the data are not sufficiently clear to inform public policy discussions concerned with the number of women in science and other professions, let alone provide a basis for specific public policies. It is entirely possible that the data will never be clear, given the inherent limitations of scientific methods applied to complex human activities. It is also possible that compelling data will point in directions politically or socially uncomfortable.

The NAS recommendations can be debated on their social and political merits alone. Scientific justifications typically are not required in matters of public policy, and scientific data often do not influence policymakers in how social goals are set or achieved. But if scientific findings are introduced as part of the debate, all the relevant data need to be considered fairly. The scientific study of sex differences in cognition will continue on many fronts, including biology, genetics, and neuroimaging, whether these approaches are endorsed by a particular NAS committee or not. Progress in neuroscience is globally competitive and inexorable. The rate of progress is tied to funding availability, and we need to ensure that funding sources are unbiased, especially when scientific questions are directed at controversial and emotional social issues like the disparities between men and women as they choose and work in various professions.

In *Why Aren't There More Women in Science?* editors Ceci and Williams concluded their summary discussion of the "knotty problem" with a quote from one chapter, "Brains, Bias, and Biology: Follow the Data."[25] It is repeated here with optimism:

> The challenge is to follow where the data lead, always cognizant of Orwellian fears and prejudiced misuse of knowledge balanced by the prospects of alleviating suffering from disorders and enhancing the quality of life for everyone. Along the way, controversy can only escalate as we constantly test new knowledge against old and comfortable ideas. This is the way science works and the way our culture evolves.[26]

Notes

1. Dr. Summers, despite this controversy, now holds a senior position in the Obama administration.

2. Halpern et al. 2007.

3. Ceci and Williams 2007.

4. Ibid., 214–17.

5. Ibid., 217–20.

6. Ibid., 220–22.

7. Ibid., 230–32.

8. Ibid., 233.

9. National Academy of Sciences et al. 2007.

10. Ibid., 2.

11. See the review of the long-running *Study of Mathematically Precocious Youth* in Lubinski and Benbow 2007, and the discussion of it in chapter 9, below.

12. Ceci and Williams 2007; Halpern et al. 2007.

13. National Academy of Sciences et al. 2007, section 2-3.

14. Ceci and Williams 2007; Halpern et al. 2007.

15. Haier and Benbow 1995.

16. Haier et al. 2004.

17. Haier et al. 2005.

18. Please note this is not the same as "intelligent design"!

19. Witelson et al. 1999.

20. Jung and Haier 2007.

21. Schmithorst and Holland 2006 and 2007.

22. National Academy of Sciences 2007 et al., 2–16.

23. Ibid., 2.

24. Ibid., 2–5.

25. Haier 2007, 113–19.

26. Ibid., 233–34.

References

Ceci, Stephen J., and Wendy M. Williams, eds. 2006. *Why Aren't More Women in Science?* Washington, D.C.: American Psychological Association.

Haier, Richard J. 2006. Brains, Bias, and Biology: Follow the Data. In *Why Aren't More Women in Science?* ed. S. J. Ceci and W. M. Williams. Washington, D.C.: American Psychological Association.

Haier, Richard J., and Camilla P. Benbow. 1995. Sex Differences and Lateralization in Temporal Lobe Glucose Metabolism During Mathematical Reasoning. *Developmental Neuropsychology* 11 (4): 405–14.

Haier, Richard J., Rex E. Jung, Ronald A. Yeo, Kevin Head, and Michael T. Alkire. 2004. Structural Brain Variation and General Intelligence. *NeuroImage* 23 (1): 425–33.

———. 2005. The Neuroanatomy of General Intelligence: Sex Matters. *NeuroImage* 25 (1): 320–27.

Halpern, Diane F., Camilla P. Benbow, David G. Geary, Ruben C. Gur, Janet Shibley Hyde, and Morton Ann Gernsbacher. 2007. The Science of Sex Differences in Science and Mathematics. *Psychological Science in the Public Interest* 8 (1): 1–51.

Jung, Rex E., and Richard J. Haier. 2007. The Parieto-Frontal Integration Theory (p-fit) of Intelligence: Converging Neuroimaging Evidence. *Behavioral and Brain Sciences* 30 (2): 135–54.

Lubinski, David, and Camilla P. Benbow. 2006. Sex Differences in Personal Attributes for the Development of Scientific Expertise. In *Why Aren't More Women in Science?* ed. S. J. Ceci and W. M. Williams, 79–100. Washington, D.C.: American Psychological Association.

National Academy of Sciences, National Academy of Engineering, and Institute of Medicine of the National Academies. 2007. *Beyond Bias and Barriers: Fulfilling the Potential of Women in Academic Science and Engineering.* Washington, D.C.: National Academies Press.

Schmithorst, Vincent J., and Scott K. Holland. 2006. Functional MRI Evidence for Disparate Developmental Processes Underlying Intelligence in Boys and Girls. *NeuroImage* 31 (3): 1366–79.

Schmithorst, Vincent J. 2007. Sex Differences in the Development of Neuroanatomical Functional Connectivity Underlying Intelligence Found Using Bayesian Connectivity Analysis. *NeuroImage* 35 (1): 406.

Witelson, Sandra F., Debra L. Kigar, and Thomas Harvey. 1999. The Exceptional Brain of Albert Einstein. *Lancet* 353 (9170): 2149–53.

9

Women, Men, and the Sciences

Jerre Levy and Doreen Kimura

The 2007 National Academy of Sciences report *Beyond Bias and Barriers: Fulfilling the Potential of Women in Academic Sciences and Engineering* attributes the greater percentage of men than women in the sciences entirely to social factors, makes multiple recommendations to increase the female percentage, and claims that standardized cognitive measures that distinguish the sexes do not predict success in scientific careers. Although we did not write this chapter as a direct response to this report, our documentation of the biological processes by which the sexes initially differentiate and then continue to diverge behaviorally provides counterarguments to many of the NAS claims. We review a wealth of evidence showing that the performance of gifted children at age twelve or thirteen on standardized cognitive tests strongly predicts the degree of success in subsequent scientific careers and even which sciences are chosen. We show that men and women arrive at their somewhat different roles and careers in life indirectly through significant prenatal genetic factors. Such factors, both early and later in life, exert physiological, primarily hormonal, influences on the brain. These influences are not limited to reproductive behaviors, but also help determine different ability patterns and lifestyle preferences in men and women. While social expectations at one time undoubtedly restricted women's choices, the evidence indicates that such choices are now largely a result of personal preferences and abilities.

Men and women are not equally represented in all occupations and professions. Both social and biological factors have, in the past, kept the number of women in professions other than teaching, nursing, and secretarial work to a small percentage. In the twentieth century, however, the lesser need for physical strength in many jobs due to mechanization, the readier availability of contraception, and the opportunities for women that were created by two world wars were associated with remarkably changed societal attitudes. While this change brought a significant degree of independence for women, it has also given rise to misleading expectations that, in a just society, men and women should be equally represented in all occupations and educational programs. Such expectations would only be realistic if the sex differences found in brain and behavior were completely uninfluenced by innate biological factors.

The unquestionable success of women in many areas of life formerly closed to them raises the question whether cultural factors alone are sufficient to explain sex differences still evident on formal cognitive tests, in choice of college majors, and in certain occupations. Although women now obtain undergraduate degrees in higher numbers than men, acquire many more advanced degrees than formerly, and have access to all professions, they still choose certain advanced science and technology fields in significantly lower numbers than do men.

This fact has recently become the occasion for much alarm concerning possible lingering discriminatory practices, and/or claims of an uncongenial environment for women in these fields.[1] Particularly vehement in such writings has been the rejection or belittling of biological factors that might contribute to the differential representation of women and men in the technical and physical science fields.

This chapter reviews the biological mechanisms that produce differences between males and females, and the development, nature, and causes of sex differences in brain, behavior, cognition, and interests.

The Biology of Sex Determination

Although most female and male babies are consistent in their chromosomal sex (XX for girls, XY for boys), their gonadal sex (ovaries or testes), their internal sex organs (such as uterus or vas deferens), and their external sex organs, each of these aspects of sexuality can be dissociated from others. Figure 9-1 summarizes the various processes in sexual differentiation, which we discuss in the following sections.

FIGURE 9-1

MECHANISMS OF SEX DETERMINATION IN THE HUMAN FETUS

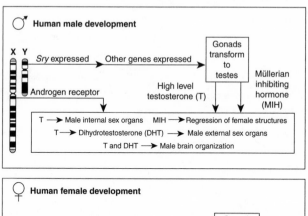

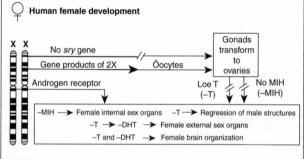

SOURCE: Authors' diagram.

From Chromosomal to Gonadal Sex. Between the sixth and seventh weeks of human embryonic life, the male-determining *sry* gene on the Y chromosome of the XY fetus is briefly expressed (turned on). This normally initiates the formation of testes from the undifferentiated gonad. It does so by regulating expression of other genes. Mutations in *sry* or other downstream genes result in complete or partial failure of the human gonads to develop (gonadal dysgenesis).[2] This contrasts with gonadal development in mice, in which the absence of *sry* in XY mice results in functional ovaries.

Also in contrast to rodents, development of the human ovaries requires the presence of two X chromosomes. Ovaries cannot develop in the human fetus if one sex chromosome is missing and cells only contain a single X chromosome (X0, a condition called Turner Syndrome). Ovarian development requires the interaction of öocytes (precursors of egg cells) with surrounding cells.[3] In all XX cells except öocytes, a majority of genes on one of the two X chromosomes are inactivated. In human XX öocytes, however, both X chromosomes remain fully active,[4] and gene products from both are necessary to maintain the cells.[5] Because gene products from a single X chromosome are insufficient, öocytes are virtually absent in human X0 gonads,[6] and the infant X0 girl is born with nonfunctional streaks instead of ovaries.

From Gonadal Sex to Genital Sex. The fetal tissues contain both male structures (Wolffian ducts) that can develop into male internal reproductive organs (epididymis, vas deferens, seminal vesicles) and female structures (Müllerian ducts) that can develop into female internal reproductive organs (uterus, fallopian tubes, upper vagina). The prostate gland and external sex organs of males (penis, scrotum) and the external sex organs of females (clitoris, vaginal lips, lower two-thirds of vagina) are formed from other fetal tissues. The developmental course is normally determined by whether the gonads are testes or ovaries.

The genetic program that controls development of female internal and external genitals is activated around the seventh gestational week in the absence of two testicular hormones, testosterone (an androgen, or "male" hormone) and Müllerian inhibiting hormone (MIH). Without these hormones, the fetal male structures (Wolffian ducts) regress and the female structures (Müllerian ducts) develop into uterus, fallopian tubes, and upper vagina. Other tissues become the external female genitalia. If the gonads fail to develop, there are no gonadal hormones in either the XY or XX fetus, and internal and external genitalia differentiate as female.

Three hormones are involved in the development of male genitalia. First, MIH from the testes causes regression of Müllerian structures and prevents formation of uterus, fallopian tubes, and upper vagina. Second, testosterone from the fetal testes induces the Wolffian ducts to form internal male organs (which would otherwise start to regress around the seventh gestational week). Third, a closely related androgen hormone, dihydrotestosterone (DHT), which is derived from testosterone in cells of target tissue, induces formation of the prostate gland and male external genitalia.

If target tissues in the fetus are deficient in the enzyme (5-alpha-reductase2) that converts testosterone to DHT, then, depending on the degree of deficiency, the child with normal testes is born with ambiguous or apparently female external genitalia. Under the action of testosterone at puberty, the penis and scrotum enlarge and take a male form. Children who have been raised as female then become males with a male gender identity.[7]

In some cases, androgen hormones are normal but have no effect on bodily tissues because there are no androgen receptors, a condition known as complete androgen insensitivity syndrome, or cAIS. The incidence of cAIS, which is due to a recessive mutation of the androgen-receptor gene on the X chromosome, is about 1 in 20,000 births of XY infants. The child has female external genitalia but neither male nor female internal sexual organs because the tissue insensitivity to androgen results in regression of the male structures, and MIH from the testes prevents development of the female structures. Although the child is a chromosomal and gonadal male, she has female external genitalia, identifies as female, and is raised as female. Estrogen from the internal testes, in the absence of androgen antagonism, induces breast development at puberty. The testes are normally removed after puberty because there is a high risk that they will become cancerous, and estrogen treatment begins.

If the fetal adrenal glands produce excess androgen (congenital adrenal hyperplasia, or CAH), a child with XX chromosomes and normal ovarian development is born with external genitalia that are ambiguously female or male to varying degrees, depending on the timing and level of fetal androgen exposure. The internal genitalia (uterus, fallopian tubes, and upper vagina, which are derived from Müllerian ducts), are normal due to absence of MIH. Some of these children are raised as males, but the large majority are raised as females. Surgical feminization of the genitalia has usually been done in infancy, with possible further surgery at a later time, but there is currently much debate about the issue and suggestions that surgery be deferred until adolescence.[8]

Sex Differences in Cognitive and Behavioral Functions

Most researchers who study sex differences in behavior make the reasonable assumption that such differences are partly shaped by our long evolutionary prehistory as hunter-gatherers.[9] During this prehistorical period men and women were not only, as they are today, different physically and physiologically, but apparently had quite divergent roles in society, which would encourage sex-typed abilities and interests. Information about the specifics of hunter-gatherer culture depends largely on fossilized bones and artifacts, but is also inferred by analogy with surviving stone-age cultures.[10] Based on existing hunter-gatherer societies, it is likely that men had primary responsibility for the manufacture of hunting weapons, and women had primary responsibility for the preparation of food and clothing and the manufacture of cooking utensils.[11]

The most salient difference in the lives of prehistoric men and women was the necessity for men to range widely in hunting or scavenging for animal food and for women to remain close to the home base to provide care and teaching of the young, to gather vegetable foods near the home site or fish in a nearby stream, and to maintain cultural continuity and social cohesion when men were absent.[12] The tendency for men to roam farther from home than women persists to the present day.[13]

Do search strategies also differ for men and women today? In hunter-gatherer culture, a hunting party in search of a herd of antelope or other animals would see trampled leaves or grass, scat, footprints, broken branches, and other signs of animal passage in proximity to the herd's location. Therefore, if one location were searched with no sign of the prey, there would be no point in seeking the herd in proximate locations, and the best strategy would be to move to distal locations (global search). The best strategy for a gatherer, however, would be to search for vegetable foods in ever-widening areas proximate to completed searches (local search). In a recent laboratory investigation of search strategies, men and women sought to find a hidden target picture among sixty-two randomly located picture cards placed face down on a large board on the floor.[14] During the first fifteen card selections (turning the cards face up), men use a global search strategy in which their typical selections are many cards distant from the immediately prior selection (scattered over the whole range of possibilities and many transition units apart). In contrast, women use a local strategy in which their typical selections are in proximity to the immediately prior selection (close together and few transition units apart). There is almost no overlap between men and women in the mean number of transition units from one selection to the next (see figure 9-2).

From this scenario, we would expect men to have evolved navigational mechanisms useful for long-distance travel, which include visualization of spatial relations and the language for communicating these relations. Such communication is descriptive of the physical environment and utilitarian in

FIGURE 9-2

SEX DIFFERENCES IN TARGET-SEARCH TASK

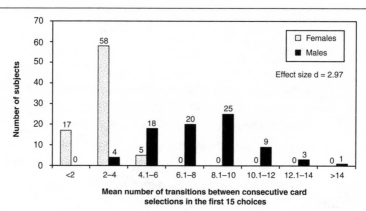

Mean number of transitions between consecutive card
selections in the first 15 choices

SOURCE: Brandner 2007 (adapted from figure 5).
NOTES: In a target-search task, women search for a target card in a large array in locations proximate
to a prior choice, but men in widely separated locations.

its aims. Women would have developed skills for navigating in nearby space with familiar landmarks, memory for object locations, skills in social interactions, emotional understanding, and language usage in service of social goals. Men would have evolved greater throwing skills than women, whereas women might be expected to have better small-amplitude motor skills. Although we have very little direct evidence of prehistoric sex differences in behavior and cognition, the evolutionary hunter-gatherer schema provides an organizing framework for understanding the origins of current sex differences and is useful in suggesting fruitful avenues for further study.

Regardless of their origins, however, differences in cognition and behavior between men and women, and boys and girls, are many and varied. Table 9-1 provides a summary of most of these differences. This is not an exhaustive list, either of abilities or sources. Nor does it imply that all other published studies comparing the sexes on the tests listed will show a sex difference. On some tests, particularly where a sex difference is typically small, there will be occasions when there is not what is called a *statistically significant* difference.[15] But it is important to note that, where an ability is listed in the table as favoring one sex or the other, only rarely do investigators observe the reverse sex difference.

TABLE 9-1

SOME HUMAN SEX DIFFERENCES IN COGNITIVE AND BEHAVIORAL FUNCTIONS

Favoring females	Favoring males
Route knowledge	*Route knowledge*
Landmark memory[1,2,3]	Route learning[3,7,8,9]
Object-location memory[4,5,6]	Geographical knowledge[10,11,12]
Perception	*Perception*
Perceptual speed[13,14,15]	Line orientation[17,18,19]
Stereoscopic fusion[16]	
Mathematics	*Mathematics*
Computation[20,21]	Mathematical reasoning[22,23,24,25]
Motor skills	*Motor skills*
Manual dexterity[26,27,28]	Throwing accuracy[32,33]
Praxis (posture and movement copy)[29,30,31]	
Verbal skills	*Spatial skills*
Linguistic usage (spelling, grammar, reading, writing)[34,35]	Disembedding[45,46]
Verbal item memory[36,37,38,39]	Spatial visualization[47,48]
Verbal fluency	Spatial rotation[49,50,51]
Begin/end letters[40,41]	Mechanical reasoning[52,53]
Making sentences[42]	
Color naming[43,44]	

Social skills
Interpreting facial expressions[54,55,56]
Social competence and problem-solving[57,58,59,60]
Emotional awareness[61,62] and eye contact[63,64]
Valuation of social goals[65,66]

SOURCES: 1. McGuinness and Sparks 1983; 2. Miller and Santoni 1986; 3. Galea and Kimura 1993; 4. Silverman and Eals 1992; 5. Dabbs et al. 1998; 6. Saucier et al. 2007; 7. Holding and Holding 1989; 8. Moffat et al. 1998; 9. Saucier et al. 2002; 10. Beatty and Troster 1987; 11. McBurney et al. 1997; 12. Beatty 2002; 13. Smith 1967; 14. Harshman et al. 1983; 15. Kimura 1994; 16. Peterson 1993; 17. Witkin 1967; 18. Robert and Harel 1996; 19. Collaer and Nelson 2002; 20. Engelhard 1990; 21. Marshall and Smith 1987; 22. Donlon 1984; 23. Benbow 1988; 24. Low and Over 1993; 25. Lubinski and Benbow 1992; 26. Tiffin 1948; 27. Hall and Kimura 1995; 28. Nicholson and Kimura 1996; 29. Ingram 1975; 30. Kimura 1997; 31. Chipman and Hampson 2006; 32. Jardine and Martin 1983; 33. Watson and Kimura 1991; 34. Feingold 1988; 35. Nowell and Hedges 1998; 36. Harshman et al. 1983; 37. Chipman and Kimura 1998; 38. Lewin et al. 2001; 39. Kimura and Clarke 2002; 40. Wilson and Vandenberg 1978; 41. Herlitz et al. 1999; 42. Ligon 1932; 43. Woodworth and Wells 1911; 44. Ekstrom et al. 1976; 45. Witkin 1950; 46. Bieri et al. 1958; 47. Bennett et al. 1961; 48. Ekstrom et al. 1976; 49. Vandenberg and Kuse 1978; 50. Masters and Sanders 1993; 51. Collins and Kimura 1997; 52. Feingold 1988; 53. Hedges and Nowell 1995; 54. Rosenthal et al. 1979; 55. Hampson et al. 2006; 56. Proverbio et al. 2007; 57. LaFreniere and Dumas 1996; 58. Bosack and Astington 1999; 59. Murphy and Ross 1987; 60. Lindeman 1991; 61. Ciarrochi et al. 2005; 62. Schirmer et al. 2005; 63. Hall 1984; 64. Podrouzk and Furrow 1988; 65. Ford 1982; 66. Jobson and Watson 1984.

We emphasize also that we are presenting *average* differences. On all tests, there is substantial overlap in the scores of men and women. So, regardless of which sex has the average or mean advantage, many individuals of the other sex will score higher than that mean. As measured by the *effect size*, which is denoted by *d*, behavior is typically less divergent between sexes than are physical differences, such as height. (An exception is the divergent search strategies of men and women shown in figure 9-2.) We obtain the effect size by subtracting the average measure (the mean) for one sex from that of the other and dividing the difference by the average within-sex standard deviation. For example, the average male height is 69.4 inches, and the average female height is 64.1 inches. The within-sex standard deviation is 2.6 inches. This gives $d = (69.4 - 64.1) / 2.6 = 2.0$ as the effect size. Effect sizes for behavioral sex differences range from small (<0.30) to large (>0.70), several of which exceed 1.0. Figure 9-3 shows the overlap between the sexes for different effect sizes.

FIGURE 9-3

OVERLAP BETWEEN TWO BELL DISTRIBUTIONS OF THE SAME VARIANCE FOR DIFFERENT EFFECT SIZES

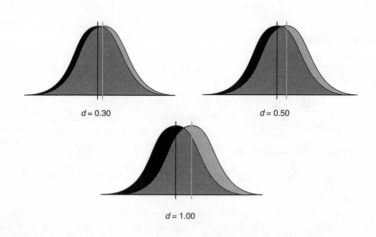

$d = 0.30$ $d = 0.50$

$d = 1.00$

SOURCE: Authors' illustration.

NOTES: The greater the offset in means between two bell curves, the greater the effect size. The effect size is denoted by *d*, which is the ratio of the difference in means to the within-group standard deviation.

Note that if male and female variances were equal, then for any effect size greater than zero, the ratio of the percentages of the favored to unfavored sex becomes progressively larger with distance above the mean (see figure 9-3). For example, if an effect size is 0.50 in favor of women, then if the variance of the sexes were equal, the percentage of women at 3.0 standard deviations or more above the female mean would be 5.8 times greater than the percentage of men. As we shall discuss, the male variance is larger than the female variance on cognitive functions. Although this may not be observed in small sample sizes, it is evident among students taking the SATs (over 1.5 million in 2007).[16] Since 15 percent more girls than boys take the SATs, boys are a more selective and restricted sample. Despite this, the variance is larger for boys than girls on all reasoning tests of the SATs, which recently have comprised critical reading, mathematics, and writing, and, in years prior to the writing test, the verbal and mathematics sections.

Even if the means of a cognitive function are identical for the two sexes, more males than females will typically be at the lower and upper extremes of the distribution. In a recent report of very large international samples, the variance of U.S. males in reading and math was, respectively, 17 percent and 19 percent greater than of U.S. females.[17] When an effect size favors males, the mean male advantage and the greater male variance operate together to increase the ratio of males to females at the upper extremes of the distribution. For example, let us assume an effect size of 0.17 in math reasoning favoring males and a 19 percent greater male than female variance. Then, at 3.0 male standard deviations or more above the male mean (top 0.13 percent of males, or 1.3 out of 1,000), there would be a little over 15 times as many men as women. At this level, women would be at or above 3.75 female standard deviations above the female mean, which means the top 0.0089 percent of women, or 8.9 women out of 100,000.

In reviewing the patterns of cognitive and behavioral sex differences described below, it will be useful to refer to table 9-1 for a summary and sources.

Route Knowledge. In learning a new route, whether in a pictorial depiction or a real-world situation, men do so in fewer trials and with fewer errors than do women. Their geographical knowledge of the world is also better. Despite that, women's recall of the presence of landmarks (structures such as buildings or bridges or other salient features) along the route is better than men's. Women apparently tend more often to use such landmarks to find their way, and also to give directions. Thus, in formal testing of people, the sexes show the preferred modes of route learning found in rodents and monkeys, with men preferring geometric modes and women preferring landmark use.

Possibly related to landmark recall is the finding that if a group of objects or pictures of objects is displayed in an array on a large piece of paper, women on average have better immediate recall of the relative location of particular objects in the array, an attribute called *object-location memory*. In contrast, women are not advantaged in their recall of location per se,[18] but only with reference to specific objects. They seem to be especially sensitive to the arrangement of objects. These findings are generally inferred to relate to the assumed evolutionary role of women in gathering food.

In a recent variation on this task, object-location memory was tested in the usual way, within personal or arm's length distance, as well as in more distant space, by projection onto a screen. There was a female advantage for object locations in personal space, but a male advantage for object locations in extrapersonal space.[19] This finding would seem to raise questions about the relevance of the standard test of object-location memory to the gathering hypothesis. The distant task was still within three meters, which would certainly be considered within "gathering" distance. Other reservations about the generality of the findings from tests of object-location memory are outlined in a meta-analysis.[20] Size or appearance of sex differences can be affected by the nature of the material or the method of scoring.

Perceptual Skills. Women are somewhat faster than men at a variety of *perceptual speed* tasks, such as finding the match for a target figure, finding words that contain a particular letter, or comparing two arrays of letters, numbers, or figures to decide whether they are identical or not. Tests of perceptual speed are essentially matching tasks in which the individual finds a stimulus identical to another. Being able to judge the presence or altered arrangement of stimuli in this fashion should enhance the capacity to detect a change in one's environment.

Men perform better at tasks requiring them to identify or match the *slope of a line*, or to set a line to the vertical or horizontal. Presumably, this could be applied to deciding whether a feature is level or an object is hanging straight. Obviously, there is a spatial element here that is not required in other matching tasks.

Social Abilities and Interests. Across a broad spectrum of nations and cultures, women are superior to men in interpreting emotional facial expressions.[21] Females also surpass males in social competence,[22] social problem-solving,[23] perception and discrimination of social information,[24] and emotional awareness.[25] Preschool girls[26] as well as adult women[27] maintain more eye contact than boys or men. The female brain responds more strongly than the male brain to emotional tones of voice.[28] Girls value social goals more than do boys.[29] Shown pairs of pictures, one of a person and the other of an object, adolescent girls spend more time looking at people than objects, and vice versa for adolescent boys.[30]

Mathematical Abilities. Males and females differ on at least two aspects of mathematical ability—calculation and mathematical reasoning. On *calculation*, where one is performing operations such as addition and multiplication, girls have a small advantage over boys. International testing in elementary and high schools finds this to be so across countries.[31]

However, when the task requires problem-solving using mathematical reasoning, males get better scores. This does not appear to be a reflection of better general reasoning ability, since strictly verbal reasoning tasks have not shown a reliable male advantage.[32] Nonetheless, most mathematical reasoning tasks on which boys excel are presented in a verbal format. It has been suggested that males are better able to transform the relevant information into a form for mathematical solution.[33]

The male advantage on math reasoning holds true even though school grades in math subjects may not differ between the sexes, or may even be higher for females (girls' grades tend to be higher in most subjects). To understand this we need to distinguish between aptitude and achievement. Grades are a measure of what has been learned in the classroom, hence achieved. Teachers aim to teach every student all the material presented, and if they are sufficiently brilliant in pedagogy, they may succeed. It is certainly never the aim to make the material so difficult that only a rare student masters it all. Aptitude tests attempt to measure abilities beyond what has been taught and are designed to test the limits. They therefore discriminate among students of different abilities better than do tests of school achievement. The distinction is seen when the same groups of boys and girls are given both kinds of tests: Girls score worse than boys only on the aptitude test.[34] A similar discrepancy between classroom achievement and aptitude tests appears also at the college level.[35] This discrepancy has sometimes been expressed as indicating that standardized aptitude tests underpredict performance of women.[36]

The male superiority in mathematical reasoning is reliable, but the effect size is quite small when assessed from unbiased samples (table 9-2). Although scores on the math section of the SATs have often been used to compare males and females in math aptitude,[37] the SATs are based on a self-selected sample in which, as was mentioned above, about 15 percent more females than males take the test. The self-selection bias overestimates the sex difference in math aptitude.[38] The bias is evident in table 9-2 when effect sizes on the SAT-Math are compared with the math section of the Preliminary Scholastic Achievement Test (PSAT-Math) from the same year. The scores on the PSAT are from high schools that required all students to take the test and are therefore unbiased. With SAT-Math omitted due to bias, the average effect size (weighted by sample size) from the years 1972–2007 is 0.17.

However, people with average mathematical ability do not become mathematicians or physical scientists, and it is at the top level of mathematical talent where males greatly outnumber females. This partly reflects the fact that even a small difference in the means of two groups has major effects at the extremes of the distribution (see figure 9-3) and partly the greater male variability,[39] which makes the disparity at the extremes much greater.

In the Study of Mathematically Precocious Youth, children selected in Project Talent (the top 3 percent of their peers) took the SATs at age twelve or thirteen, when they were in the seventh or eighth grade. Among those who scored 700 or above[40] on the math SAT (1 in 10,000 at this young age), there were eleven times as many boys as girls.[41] An even greater disparity is seen at the high school level in the International Mathematics Olympiads (IMO). Between 1964 and 2008, thirty-one participants won three or more gold medals. Only one is a girl, but she achieved perfect scores in more than one year, a feat accomplished by only four of the thirty boys.[42] In a very advanced mathematics competition for college undergraduates in North America, the Putnam competition,[43] a much higher percentage of males than

TABLE 9-2

EFFECT SIZES FOR MATHEMATICAL REASONING IN
UNBIASED SAMPLES AND ON SAT-MATH

Study and databases examined	Sample size	Test	Effect size
Nowell and Hedges 1998[1]			
1960 Project Talent (PT)	2,807	Arithmetic reasoning	0.31
1965 Equality of Educational Opportunity (EEO)	48,000	Understanding and applying mathematical concepts	0.31
1972 National Longitudinal Study of the High School Class of 1972 (NLS)	8,430	Are two quantities equal or is one greater or impossible to tell?	0.24
1980 High School and Beyond (HSB)	12,534	Are two quantities equal or is one greater or impossible to tell?	0.22
1982 High School and Beyond (HSB)	11,623	Are two quantities equal or is one greater or impossible to tell?	0.09
Feingold 1988			
1974 High school juniors (national standardization sample)	17,658	PSAT-Math	0.17
1974 High school juniors (self-selected)[2]	*340,000*	*SAT-Math*	*0.42*
1983 High school juniors (national standardization sample)	25,316	PSAT-Math	0.12
1983 High school juniors (self-selected)[2]	*340,000*	*SAT-Math*	*0.42*
College Boards 2007			
2007 PSAT (10th grade)—D.C., Georgia, Maine, Nevada (100% student participation)	145,791	PSAT-Math	0.18
2007 SAT (10th grade)—D.C., Georgia, Maine, Nevada (self-selected)[2]	*85,860*	*SAT-Math*	*0.28*
2007 All students who took the SATs (self-selected)[2]	*1,494,531*	*SAT-Math*	*0.30*

NOTES: 1. The 1980 National Longitudinal Study of Youth (NLSY), the 1992 National Education Longitudinal Study of the Eighth Grade Class of 1988 (NELS), and the multiyear National Assessment of Educational Progress (NAEP) are omitted because they do not test math aptitude. 2. The SATs are based on a self-selected sample in which a more select group of males than females takes the test. The self-selection bias overestimates the sex difference in math aptitude (Lewis and Willingham 1995). Studies in italics suffer from self-selection bias.

females are among the top prize winners. In the years 2003–7, twenty-seven top prizes of $2,500 were awarded to nineteen different students (several of whom won in multiple years). All are male except one, who was a top winner in both her freshman (2003) and sophomore (2004) years. During her high school years, she also won two gold medals in the IMO. Three of the eighteen male Putnam winners were among the thirty high school students who had won three or more gold medals in the IMO since 1964. Finally, the Fields Medal, which is the Nobel Prize of mathematics, has been awarded to forty-eight people since it was first given in 1936, none of whom has been a woman. This obviously does not imply that no woman can do the caliber of work worthy of a Fields Medal. Emmy Noether (1882–1935), whose contributions permeate all fields of algebra, would probably have been a recipient had the medal been awarded during her lifetime. But women of such extreme mathematical ability are a very small minority among people gifted with mathematical genius.

Motor Skills. Unquestionably, men are, on average, physically stronger than women. This probably contributes to the fact that they are faster on some repetitive motor tasks, and it may help explain their superiority in some sports. Men are also more accurate in throwing a missile at a target and in catching an object, strength being a less critical factor in the latter. Catching requires prediction of and appropriate motor response toward a moving target, which perhaps relates to a male advantage in predicting the motion of dynamic spatial displays on a video screen, especially for velocity judgments.[44] Men's advantage in predicting the time at which two objects will make contact is also more marked for far than for near space.[45] This is in agreement with sex differences in favored space on other tasks. The sex difference in throwing and catching accuracy is as large as that on the most differentiated paper-and-pencil cognitive tasks and is apparently not simply accounted for by sports experience.[46] It may be related to the distinctive visual and spatial nature of motion directed at a point in external space.

In contrast to the male superiority on motor skills related to extrapersonal space, women tend to be superior on motor tasks within personal, or arm's length, space. Their better dexterity in, for example, handling and placing small parts has been repeatedly demonstrated. Even when dexterity is not a critical component, however, as in copying or learning a series of hand and arm movements, women perform better than men. The fact that women are more accurate at making multiple serial movements within personal space would be conducive to performance of many tasks that are demanding of motor programming but not necessarily of spatial analysis. These would include laboratory skills in the biomedical sciences, as well as certain traditional female activities (for example, knitting or sewing). Women rely less on vision in doing manual sequencing tasks than men do.[47]

Verbal Abilities. The general belief that women surpass men on all or most verbal skills is not a valid one. On the verbal parts of adult intelligence tests, there is either no significant difference between the sexes, or men may score slightly higher.[48] Even on vocabulary, where we might expect women to excel because they read more than men,[49] a difference does not appear. Nevertheless, females do excel in several verbal skills. They are typically better at spelling, formal grammar, and reading, well into the late high school years. Females are also reported to be better at writing in the twelfth grade and on the recently introduced SAT-Writing component, which is also taken at the end of high school. In part, the multiple-choice section of the writing test assesses grammatical understanding and usage, although not spelling.[50]

Women are also better at a skill called *verbal fluency*. This doesn't mean that females speak more fluently. It refers, rather, to the speed with which they can generate words. On one commonly used test, people say as many words as possible within a time limit that begin with a particular letter. Women typically produce more words than men. Women also surpass men in the rate of generating color names of circular patches on a sheet.

The verbal task that so far has yielded the largest female advantage, however, is *verbal memory*. The advantage holds true for recalling both lists of unconnected words and meaningful material, such as paragraphs. It applies to both abstract and concrete words, and it extends also to recall of objects that are easily named. A good verbal memory is essential for lawyers and editors, and probably for many administrative jobs. Women might be more likely to remember conversations accurately, and, to men's chagrin, anecdotal evidence favors that view.

Good verbal memory might also be expected to enhance landmark recall in route learning, though we have as yet no explicit evidence to that effect. It should enhance the ability to itemize objects in the immediate environment, that is, to make implicit lists.[51] Coupled with better perceptual matching ability and better object-location memory, it could enable women to maintain a more precise inventory of things in their environment and to remember where the objects are.

Spatial Abilities. It is widely accepted that males outperform females on a variety of tasks that require spatial ability of some kind. We have already mentioned route learning and geographical knowledge, as well as ability to identify line orientation. Some less commonly performed tasks also show a male advantage.

The kind of task that gives the largest male advantage requires what is called *mental rotation*. To solve the problem, one must correct for the orientation of an object or imagine how it would look if rotated in one plane. Commonly used figures depicting three-dimensional objects that are used to test for this ability are shown in the upper portion of figure 9-4.[52] They are composed of several segments and are shown as rotated around a central axis. Mental rotation tests of this kind yield the largest sex difference of any paper-and-pencil test. This finding has given rise to the belief that a three-dimensional element is necessary to demonstrate a large sex difference.[53] It is more likely, however, that the difficulty of the task is the critical factor, since a very difficult two-dimensional rotation test (bottom portion, figure 9-4) has shown an equivalently large sex difference.[54]

For both men and women, the ability to perform mental rotation is significantly related to ability to navigate a route,[55] but especially when using geometric cues.[56] One might speculate that in navigating long distances, an ability to recognize a particular scene from different angles would contribute to orienting one's self, and thus finding one's way. Males also have an advantage on spatial tests that require mental folding, assembling, or other forms of spatial visualization. These tests involve imagining what happens when a figure or figures are manipulated in some way, such as by folding, or by assembling several figures into one. As previously noted, there is a reliable male advantage in adjusting a line to the absolute vertical or horizontal within a tilted frame, which is called *field independence*.[57] Male advantages in spatial visualization and field independence are consistent, though generally smaller than those on the most difficult mental rotation

FIGURE 9-4

DIFFICULT 3D AND 2D MENTAL ROTATION TASKS

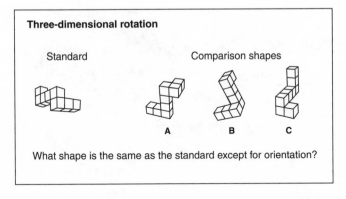

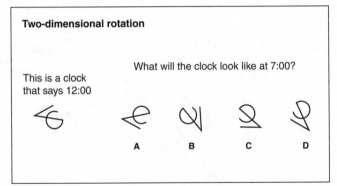

SOURCES: Top panel adapted from Vandeberg and Kuse 1978; bottom panel adapted from Collins and Kimura 1997.
NOTES: The answer to the top and bottom items is C in both cases.

tasks. Another sex-sensitive test, requiring the identification of one figure hidden within a complex figure (a skill called *disembedding*), also shows a smaller sex difference.

All these abilities are related to the large male superiority in mechanical reasoning. Accurate responses to test items like those shown in figure 9-5 may require spatial visualization or rotation, intuitive geometry or physics, or an understanding of the principles of simple machines (such as levers or

FIGURE 9-5

TYPICAL ITEMS FROM MECHANICAL REASONING TASKS

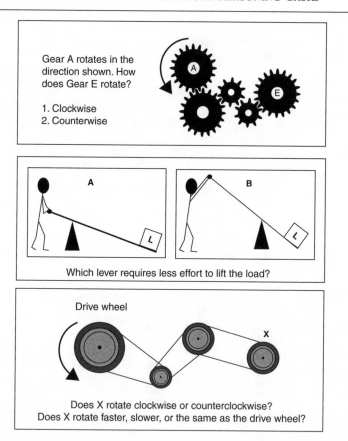

SOURCE: The authors created these three items based on similar ones in many tests of mechanical reasoning.
NOTES: The answers to the top and middle items are 2 and B, respectively. The answers to the bottom item are counterclockwise and faster.

pulleys). One would expect these kinds of abilities to be important for various constructional, mechanical, and engineering skills.

How large are the differences just described? The convention is to call an effect size of roughly 0.7 to 1.0 or greater "large." An effect size of less than 0.3 is considered small, and the rest are intermediate. The effect sizes on difficult mental rotation tasks and mechanical reasoning (figures 9-4 and 9-5) vary around 1.0 or greater. Although the effect size on disembedding is much smaller (often under 0.3), the effect size in representing a horizontal water line in a tilted frame[58] is considerably larger (around 0.55).

Cross-Generational and Cross-Cultural Comparisons

It is sometimes stated that if biological factors influence cognitive abilities, they must be "immutable."[59] We know of no biological scientist who makes such a claim for behavior. The dynamic nature of biology makes interaction of biogenic and environmental influences unavoidable. Indeed, it has been suggested that cultural factors favor certain traits over others and thereby select genes underlying the more favored traits. This would increase the frequencies of favored traits and associated genes in the next generation, which could influence cultural evolution and, in turn, subsequent biological evolution.

All cognitive sex differences partly reflect social and educational factors and would be expected to decline in magnitude over time as these factors become more equalized between sexes. However, to the extent that these differences are also influenced by biology, they would not entirely disappear even if expectations and sex roles were identical. Similarly, while we would expect to find variations across cultures in the magnitude of sex differences depending on the differences or similarities in sex roles and opportunities, biological influences predict the same direction of sex differences across cultures. This would not in itself prove biological causation, but a failure to find any cross-cultural consistency would be strong evidence against biological influences.

Across Generations. The question has arisen whether sex differences in cognition have diminished over the past few decades during which men's and women's educational and work experiences have become more alike. In a 1988 study that addresses this question, Feingold analyzed scores for very large samples of American high school students on the Differential Aptitude Test (DAT).[60] He assessed changes in the size of the sex differences across four periods during 1947–80 for grades 8–12. Although the effect sizes on spelling, language (grammar), and clerical skill, which favor girls, and on mechanical reasoning and spatial relations, which favor boys, declined with time, none approached zero. Because girls' and boys' scores are not given separately, it is difficult to pinpoint the precise source of the apparent diminution in sex differences. The text suggests that boys gained more than girls on the female-favoring tests, while girls gained more than boys on the male-favoring tests.

There were no reliable sex differences on the numerical ability test, which emphasizes school achievement, but on other non-DAT math tests (PSAT and SAT), which emphasize math aptitude, there was, during the same time period, the usual male advantage. As discussed, the effect size on SAT-Math is an overestimate of the sex difference (see table 9-2) due to selection bias, in which males are a more highly selected sample than females. To the extent that the same selection bias operates over time, which seems to be the case from the data in table 9-2, this would reduce a possible decline with time in the effect size of mathematical aptitude. Feingold reported that "on PSAT-Mathematical, the gender difference among juniors (the only students tested four times) decreased from .34 in 1960 to .12 in 1983, a 65% decline."[61] Nowell and Hedges also reported a decline over time in the effect size for mathematics, as well as for science knowledge.[62]

Criticisms of Feingold's study of DAT scores, acknowledged and discussed by him,[63] include suggestions that since different individuals were tested in the four time periods, some of the apparent reduction in the sex difference may have been due to the differing cohorts being tested. For example, if the number or composition of males and females was, for various reasons, different from one cohort to the next, this might have affected scores. Nowell and Hedges found no decline over years on most cognitive measures (exceptions were mathematical aptitude and science knowledge) and suggested that Feingold's sample of students may not have been as representative as theirs.[64] Other suggestions that test items favoring one sex or the other were dropped in later years might also be valid.

The effect sizes still remained above 0.40 on language tests (favoring females) and at 0.76 on mechanical reasoning (favoring males) in 1980, the latest year Feingold reported. Feingold suggested that biological influences might be greater for some abilities than others, and might therefore continue to show substantial sex differences, despite the more equal access of males and females to relevant experiences.[65]

The much-used spatial test of mental rotation[66] described earlier has shown a large male advantage that did not systematically change among college students between 1975 and 1992.[67] More recent studies have shown similarly large effect sizes, hovering around 1.0.[68] On this type of test, therefore, as of 2006, there had been no decline in the sex difference within more than thirty years.

The view that life experience is the major cause of sex differences in ability has been inferred by some authors from a relation between past life activities and current abilities, specifically for mathematical and spatial ability, and particularly for the latter.[69] In one study, the investigators report that experiences in spatially oriented activities are weakly related to scores on a spatial test that requires choosing which of four depicted boxes can be made by folding a two-dimensional figure. A more recent study compared college sailors, crew members, and the general student body on a classical test of mental rotation and found that sailors have higher scores than others. In an East African study, researchers made naturalistic observations of the distances from home traveled by girls and boys and compared these to the ability to copy block patterns and geometric figures. Boys travel greater distances and score higher on the block patterns and figures.[70]

Such findings in themselves do not constitute evidence that past activities *determine* present abilities. People who are spatially clever most probably engage in spatially demanding activities all their lives, simply because they can. The ability may determine the experience, rather than the reverse. Boys and girls have different hormonal makeups even before birth, and we will later discuss probable hormonal contributions to various cognitive skills and behaviors that are sexually dimorphic.

A similar correlational fallacy is that young girls and boys behave differently because their parents treat them differently. An equally plausible explanation is that parents treat boys and girls differently because the developing infants behave differently. If we know only that two facts are associated (boys and girls behave differently; parents treat boys and girls differently), we must be cautious about inferring a causal connection.

If experience influences sex differences in spatial cognition, then specific training in spatial tasks might reduce those differences. A review of pertinent studies confirmed a male advantage in spatial cognition and showed that both females and males improve with practice.[71] Contrary to the authors' expectations, however, the improvement with practice is no greater for females than males, and the male advantage does not decline. Another study found that the sex difference on a mental rotation task remains unchanged even after long-term, intensive practice.[72] To the extent that the abilities measured by such tests are relevant to fields such as engineering, architecture, geometric reasoning, and general spatial problem-solving, it is unlikely that training alone can equalize the numbers of men and women in spatially demanding occupations and professions.

Across Cultures. Another way of addressing the significance of life experience for the emergence of cognitive sex differences is to compare cultures that vary in sex roles. One study found that students from both Japanese and American private prep schools manifest typical sex differences on word fluency, verbal memory, perceptual speed, and mental rotation.[73] A South African study using a large sample found a similar pattern in blacks and whites, but for East Indians only tests favoring males showed a significant difference.[74] These sex differences appear despite some differences among ethnic groups in overall scores.

Tests that favor males have been more extensively studied cross-culturally than those that favor females. In investigations of people in China and America, males perform better than females on two map-reading tests but not on a third that uses street names.[75] Rotation and other spatial tests yield sex differences in some African and East Indian groups.[76] In a water-level task of horizontality, Indian boys are better than girls, but only at older ages.[77] Equivalently large male advantages for mental rotation tests have been found across Canadian and Japanese,[78] American and Chinese,[79] and Canadian, German, and Japanese groups.[80] These sex differences again appear despite some differences across ethnic groups in total score.

A reported failure to find a sex difference in either Ecuadorean or Pakistani schoolchildren on presumptively spatial tests was imputed to the traditional role of women in these cultures in weaving and constructional work.[81] However, the tasks employed by the researchers have never been demonstrated to favor males in any nation, and it is quite possible that North American boys and girls would also perform equally.

Somewhat more convincing is the claim that typical sex differences on spatial tasks do not appear in the Inuit (Eskimo). Though very little detail is given, two studies report that on the Witkin[82] disembedding task, no sex difference is seen in the Inuit sample; yet in one of the studies, a sex difference appeared in both a Scottish and an African group.[83] Both studies attributed the absence of male superiority to the lack of male social dominance in the Inuit as compared to other ethnic groups, although there is no evidence that social dominance is related to spatial cognition. A more probable explanation is that survival in the nearly featureless landscape of the Inuit demands good visual perceptual skills in women as much as in men. We know of no studies on the Inuit comparing sexes on difficult mental rotation tasks, nor on targeting, which yield the largest male advantages. Disembedding tasks yield a less robust male advantage than other spatial tasks and may be too insensitive to reveal a reliable sex difference in the Inuit. Although Canadian Caucasian samples on one version of a disembedding task[84] showed a slight male advantage, it was not statistically reliable.[85]

On mathematical tasks, both the male advantage for math reasoning and the female advantage for computation have been found in American blacks, whites, Hispanics, and Asians,[86] and in students from the United States and Thailand.[87] The male advantage on math reasoning has been reported for school samples in the United States, China, and Japan,[88] and in American Asian and Caucasian samples.[89] The male advantage for SAT-Math 2007 appears in all ethnic groups, but the effect size varies from a low of 0.14 to 0.35. In all ethnic groups, more females than males take the SATs, and, as mentioned previously, such selection bias overestimates the male advantage in the population.

In summary of this section, it appears that the magnitude of sex differences in mathematical aptitude and science knowledge has declined over the past three or so decades, but a reliable male advantage remains. There is no trend over time and no effect of training on the male advantage in mental rotation. Whether sex differences in other cognitive abilities are shrinking over time is unclear from current studies, because some report a decline and others find no change. Although sex differences that favor males or females vary in magnitude across cultures and ethnic groups, the direction of these differences is generally the same. Such observations are consistent with the view that cognitive or behavioral sex differences reflect both sociocultural and biological influences.

Behavioral Sex Differences in Infants

Most of the cognitive characteristics that differ in men and women cannot be assessed in infants because the cognitive processes are not developed and will not develop for some years. However, even newborns can be evaluated for social interests. If the female advantage in social cognition and the greater female than male interest in the social world are due partly to biological factors, then sex differences in social behavior might emerge in infancy.

Infant girls prefer faces over other visual objects. Given a choice between a mobile and a face, newborn girls, but not boys, more frequently look at the face and spend more time looking at the face.[90] The face stimulus in this study was one of the female researchers, who maintained a neutral expression during a trial. It is possible that unconscious experimenter bias may have resulted in a friendlier face being presented to girls than boys. However, other studies of infants observe the same female preference for faces over nonfaces. Female infants two to six months of age spend more time visually tracking a face-like stimulus than do boys and less time tracking either of two nonface stimuli.[91] Boys track all stimuli equally often. At twelve months of age, girls spend more time looking at silent video clips of two people talking or the head and shoulders of a male reading aloud than at cars racing at high speed or one car with its windshield wipers moving. Tests with boys yield the opposite results.[92] The greater attention given to faces by infant girls is consistent with an adult female advantage in understanding emotional and communicative facial expressions. Similarly, the greater attention given to objects by boys than by girls is consistent with an adult male advantage in understanding mechanical or spatial relations.

Girls at twelve months of age are superior to boys in engaging in social interactions. In particular, they are more skilled in using nonverbal behaviors to elicit another's attention to a toy, in eliciting aid to obtain a toy, and in engaging in playful turn-taking interactions with others.[93] Infant girls also manifest a greater interest in and responsiveness to social interactions than do boys. In a free-play situation between mothers and infants at ages six, nine, and twelve months, mothers initiate social interactions with male and female babies equally often and respond equally often to male and female infants' initiations. However, girls initiate more interactions than do boys and respond more frequently to their mothers' vocal initiations.[94] If the sex differences were entirely due to socialization, they should increase with the age of the baby. They are, however, just as large at six as at twelve months.

The greater social orientation of girls than boys is also manifested in their greater eye contact. In newborns[95] and twelve-month-olds,[96] females make more eye contact than males, just as in preschoolers and adults. Lutchmaya and her colleagues report that the relation between amount of eye contact at twelve months of age and level of fetal testosterone (assessed from amniotic fluid taken at amniocentesis) is curvilinear.[97] Eye contact is greatest at the lowest levels of testosterone (in females) and at the highest levels (at the high end of males) and least at intermediate levels (at the low end of males). A very similar curvilinear relation is seen in adults between circulating testosterone levels and behavior: Male behavior departs most from female behavior when testosterone is at the low end of the male range.[98]

The infant studies suggest an inherently stronger social orientation in girls than boys, who display as much or more interest in objects or the physical world as in people. It is not unreasonable to suggest that throughout development, girls on average would select environmental experiences that nourish their social interests and skills, whereas boys, on average, to a greater extent than girls, would select those that nourish their interests in the physical world and their skills in spatial and mechanical understanding. Such transactions between biological tendencies and experiences that enhance their development would result in progressive cognitive differentiation of the two sexes.

Biological Influences in Early Life on Behavioral Sex Differences

The brain has a sexual identity at birth, which is dramatically highlighted in the case of an infant, Bruce Reimer, who was born a boy, lost his penis at the age of eight months from horribly botched surgery, and was reared as a girl at the recommendation of John Money, a presumed expert at Johns Hopkins University. Money assured the parents that infants are psychosexually neutral and the child would be a normally adapted little girl. The infant Bruce became Brenda. The testes were removed, and surgery feminized the genitalia. But Bruce/Brenda never accepted a female identity, rejected female clothes, toys, and activities, and at age fourteen adopted a male identity and chose the name

David. When David was fifteen, his parents finally told him the truth of his history. He said, "'Suddenly it all made sense why I felt the way I did. I wasn't some sort of weirdo. I wasn't crazy.'"[99]

Although David Reimer subsequently underwent reconstructive surgery to create male genitalia and married, the tragedy of his childhood was never resolved. He killed himself at the age of thirty-eight. According to his biographer, John Colapinto, "The real mystery was how he managed to stay alive for 38 years, given the physical and mental torments he suffered in childhood and that haunted him the rest of his life. I'd argue that a less courageous person than David would have put an end to things long ago."[100] A denial of inherent differences in the brains and behaviors of males and females gave Bruce/Brenda/David Reimer a tortured childhood, adolescence, and adulthood, and finally cost him his life at a young age.

Early Hormonal Influences. Male and female mammals and birds typically differ not only in mating and parental behaviors, but also in interests, activities, and skills or abilities that have no direct relation to reproductive behaviors. Certain of these differences are remarkably similar in human and nonhuman primates, which suggests commonalities of underlying mechanisms. Boys[101] and juvenile male monkeys[102] engage in more rough-and-tumble play than girls and juvenile female monkeys. Girls[103] and female monkeys[104] display more interest in infants than do boys and male monkeys. Toys that are more attractive to girls than boys (dolls, pots) are also more attractive to female than male monkeys, whereas toys that are more attractive to boys than girls (balls, toy vehicles) are also more attractive to male than female monkeys.[105]

Studies of nonhuman animals, including primates, establish the central role of gonadal hormones during fetal or infant life in producing sex differences in the brain and behavior. Androgen treatment of females in the prenatal or early postnatal period increases male-typical and decreases female-typical behaviors, whereas castration of males at birth or treatment with anti-androgens has the reverse effects.[106]

The primate neocortex, the highest region of the brain, is very poorly developed at birth and undergoes rapid maturational changes in the first six months of life in monkeys, apes, and people. Androgen speeds the maturation of some neocortical regions and slows the maturation of others. On a task that depends on a neocortical region that matures earlier in males (the orbital frontal cortex, just behind the eyes), infant male rhesus monkeys perform better than infant females, a sex difference that disappears when the region has matured in females. However, infant females treated with androgen perform as well as infant males and significantly better than untreated infant females.[107] On a task that depends on a brain region that matures earlier in females than males (the inferior temporal cortex, next to the ears), infant female rhesus monkeys perform better than infant males, a sex difference that disappears by six months of age when the male region has matured. Male infants castrated at birth perform as well as female infants, whereas infant females treated with DHT perform as poorly as infant males.[108] Due to effects of androgen on the infant brain, the developmental pattern of the neocortex differs for male and female infants.

Vocalizations appear earlier in female than male infant monkeys[109] and differ between the sexes in their characteristics.[110] Male infants use more high-energy screams and female infants use more low-energy coos when separated from or rejected by their mothers. Mothers respond more to male than female distress calls, quite possibly because high-energy screams draw greater attention than low-energy coos. When females are treated prenatally with androgen, their infant vocalizations become completely masculinized. When males are treated prenatally with an anti-androgen, their infant vocalizations are partially feminized.

These studies of infant monkeys demonstrate that androgen hormones act in the prenatal or infant periods to speed or retard development of the cerebral cortex, depending on the brain region, and to masculinize communicative vocalizations. Differences between the sexes in which regions of the cortex mature earlier are likely to influence which aspects of the world are perceived and which elicit attention, interest, and play behavior. The resulting differences in experiences would affect subsequent brain development.

Evidence that androgen in early development has similar effects on human behavior comes from several sources (see tables 9-3 and 9-4, below). The case of David Reimer shows that a chromosomal male whose fetal brain is organized by androgen from the testes retains a male identity, interests, and activities even when reared as a girl with female external genitalia and no knowledge of his male history.[111]

Male interests and play activities also characterize children with XY chromosomes and normal fetal testosterone who were born without a penis. Reiner and Gearhart describe sixteen children with this very rare condition. Two are being reared as boys and have a strong male identity and male-typical behaviors. The other fourteen were surgically feminized in infancy and are being reared as girls. Despite sex of rearing, six of the fourteen have spontaneously adopted a male sexual identity. Two others declared a male identity when they were initially examined, but at the latest examination, their sexual identity was unclear. Another child refused to discuss his/her sexual identity. The remaining five children have a female identity, but one expresses a wish to be a boy. Regardless of sexual identity, all the children have male-typical toy preferences and play behaviors.[112] Thus, we may conclude, prenatal exposure to male levels of androgen overrides the effects of female genitalia and rearing in governing activities and interests. Although male sexual identity or confusion over sexual identity characterizes a small majority of these children, some accept a female identity, congruent with rearing. This demonstrates, as Berenbaum and Sandberg have noted, the complexity of factors that determine sexual identity, which is separate from interests and activities.[113]

In contrast to the masculinizing effects of prenatal androgen on behavior, a lack of prenatal androgen or tissue insensitivity to its normal effects results in feminine behavior in chromosomal males. Chromosomal males with androgen insensitivity are born with female external genitalia, have a female gender identity, and display feminine interests and activities that are just as pronounced as in typical girls and women with XX chromosomes.[114]

The level of prenatal androgen exposure also affects the brains and behavior of chromosomal females with normal ovarian development. First, although even the highest level of maternal androgen during pregnancy is very much less than produced by the testes of fetal males, it influences brain organization of female fetuses and the subsequent behavior of girls and women. The higher the level of maternal androgen to which females are exposed in fetal life, the less feminine and the more masculine are their sex-typed behavioral characteristics.[115] Second, excessive prenatal androgen exposure occurs in congenital adrenal hyperplasia, in which the fetal adrenal glands secrete high levels of androgen and fail to synthesize cortisol and other critical hormones. Because the condition can be fatal, medical treatment for CAH is typically initiated at birth, which replaces cortisol with drugs such as hydrocortisone and reduces androgen in girls to normal or below-normal female levels. However, the exposure of the fetal brain in CAH females to levels of androgen much greater than in their healthy sisters or other female relatives results in greater masculine and less feminine sex-typed interests (for example, greater interest in electronics and less in fashion), vocational goals (engineer versus X-ray technician), toy choices (trucks versus dolls), personality characteristics (more aggressive, less empathetic, less interest in infants), play activities (more rough-and-tumble play), and characteristics of preschool drawings (more vehicles, fewer flowers, less pink, more arrangements in piles than rows).[116]

Table 9-3 summarizes studies of the effects of prenatal hormonal exposure on activities, interests, personality and vocational goals.[117]

TABLE 9-3

EFFECTS OF ANDROGEN IN EARLY HUMAN DEVELOPMENT ON INTERESTS, PERSONALITY TRAITS, AND ACTIVITIES

Prenatal environment	Genitalia and rearing	Behavior
Normal male androgens, XY chromosomes (David Reimer).	Female external genitalia. Reared as female.	Masculine interests and activities.[1,2]
Normal male androgens, XY chromosomes (testes but no penis).	Genital feminization in infancy. Reared as female.	Masculine toy preferences, play behaviors, playmates.[3]
Complete androgen insensitivity syndrome (cAIS), XY chromosomes.	Female external genitalia. Reared as female.	Feminine interests and activities as prevalent as in other girls and women.[4,5,6,7]
Variations in fetal exposure to maternal androgen, XX chromosomes.	Female genitalia. Reared as female	Sex-typed behaviors more feminine or masculine with low or high exposure, respectively.[8,9,10]
Congenital adrenal hyperplasia (CAH; excess prenatal androgen), XX chromosomes.	Ambiguous genitalia at birth, surgically corrected to female. Reared as female.	Reduced feminine and increased masculine sex-typed behaviors.[11,12,13,14,15,16,17,18,19]

SOURCES: 1. Diamond and Sigmundsen 1997; 2. Colapinto 2006; 3. Reiner and Gearhart 2004; 4. Hines et al. 2003; 5. Melo et al. 2003; 6. Jürgensen et al. 2007; 7. Wisniewski et al. 2000; 8. Udry et al. 1995; 9. Ehrhardt et al. 1977; 10. Hines et al. 2002; 11. Berenbaum 1999; 12. Berenbaum and Bryk 2008; 13. Berenbaum and Snyder 1995; 14. Cohen-Bendahan et al. 2005; 15. Hines 2004; 16. Hines et al. 2004; 17. Iijima et al. 2001; 18. Nordenstrom et al. 2002; 19. Pasterski et al. 2007.

Girls and women with CAH also differ from their healthy sisters or other control females in skills or abilities for which there is an average male advantage. They are much more accurate than control females in throwing darts or balls at a target and are as accurate as typical males in dart-throwing.[118] In those spatial skills for which there is an average male superiority, most studies report enhanced ability in CAH compared to control females, although findings vary in the magnitude of enhancement and in whether an outcome is statistically significant[119] or not.[120] In their review and analysis, Puts and colleagues have concluded that CAH females have better spatial abilities than control females.[121]

Table 9-4 summarizes evidence that prenatal androgen influences skills and cognitive abilities that differ between the sexes.

TABLE 9-4

EFFECTS OF ANDROGEN IN EARLY HUMAN
DEVELOPMENT ON SKILLS OR ABILITIES

Prenatal environment	Genitalia and rearing	Skill or cognition
Complete androgen insensitivity syndrome (cAIS), XY chromosomes.	Female external genitalia Reared as female.	Normal verbal cognition but low perceptual/spatial cognition.[1*,2*,3]
Congenital adrenal hyperplasia (CAH; excess prenatal androgen), XX chromosomes.	Ambiguous genitalia at birth, surgically corrected to female. Reared as female.	Superior targeting accuracy (dart- or ball-throwing) compared to unaffected female relatives. Equal to healthy males in dart-throwing.[4*] Better ability than control females on spatial tasks for which males are superior to females.[4,5*,6*,7*,8*,9,10,11] This relation was not observed in one study.[12]

SOURCES: 1. Imperato-McGinley et al. 1991; 2. Masica et al. 1969; 3. Spellacy et al. 1965, a single case report; 4. Hines et al. 2003; 5. Berenbaum et al. 2006; 6. Hampson et al. 1998; 7. Resnick et al. 1986; 8. Mueller et al. 2008; 9. Baker and Ehrhardt 1974; 10. Helleday et al. 1994, in a subset of CAH and control females matched for general intelligence; 11. Malouf 2006 et al., in CAH women without a salt-losing disorder; 12. Ibid., in salt-losing CAH women.
NOTE: * = Statistically significant difference.

Although the body of research suggests that spatial abilities are enhanced in CAH compared to unaffected females, the variations among studies in the degree of enhancement and in statistical significance contrast with the consistently large effects of CAH on sex-typed activities, attitudes, interests, personality variables, and targeting ability. Whether these characteristics are more susceptible than spatial ability to the masculinizing effects of prenatal androgen is an open question. Alternative possibilities are that individuals with CAH are highly variable in spatial cognition due to variability in disorders of brain function associated with their illness or to variability in the timing and degree of fetal androgen exposure, which affects some behaviors more than others.

Direct Effects of Genes on the Brain and Behavior. Although most genetic sex differences in the brain and behavior are mediated by gonadal hormones, some reflect the direct actions of genes.[122] Differences in genes on the sex chromosomes influence development even before the embryonic gonads begin to differentiate into male testes or female ovaries. For example, embryonic brain cells cultured in a petri dish (in vitro) in the absence of any hormonal influence develop differently, depending on whether they are XX or XY cells. Studies of postmortem human brains reveal genes on the sex chromosomes that are expressed only in one sex, or at a higher level in one sex than the other.[123]

Differences in XY and XX cells arise through a variety of mechanisms, which are outlined in table 9-5, as follows:

- Genes unique to the Y chromosome can only affect males.

- Recessive genes on the X chromosome that have no partner on the Y chromosome (no Y homologue) produce the recessive trait in males but not in females, unless both X chromosomes carry the recessive gene.

- Although one random X chromosome in each cell of females is partly inactivated early in development, certain genes escape inactivation. The escapees include not only genes on the tips that are identical to genes on the tips of the Y chromosome, but genes on other regions of the X that have no Y homologue. Since such genes are expressed on both X chromosomes in XX cells but only on the single X chromosome of XY cells, the dosage is greater in female than male cells.

- Even when a gene on the X chromosome is identical to a gene on the Y chromosome, genes on the X and Y may be expressed at different levels, with different timing in development, and in different tissue regions due to regulatory genes on the X and Y.

- Certain genes are imprinted with their parental origin and are only expressed from a single parent. Imprinted genes on the X chromosome have been identified in animals and are implied by behavioral studies of girls or women with Turner syndrome (TS), who have only a single sex chromosome (X0 instead of XX).

In the X0 condition, girls and women have normal verbal cognition and reduced perceptual or spatial cognition. A majority also have impaired social skills, although a minority have good social skills. The social deficits are almost certainly related to the high rate of autism (up to 5 percent) and autistic spectrum disorders (greater than 25 percent) in X0 females.[124]

Girls with TS whose single X is of paternal origin (X^P), which only daughters inherit, have better verbal cognition, better social cognition, a lower risk of autism, and a greater capacity for behavioral inhibition than those whose single X is of maternal origin (X^m), which both sons and daughters inherit. The differences between X^P0 and X^m0 girls suggest that genes on the X chromosome that enhance social cognition, reduce the risk of autism, and increase the capacity for behavioral inhibition are imprinted and differentially expressed on the paternal X chromosome.[125]

Girls and women with TS who carry the maternal or paternal X chromosome also differ in retention of visual and verbal material over a forty-minute delay period. When the effects of individual differences in baseline scores (for copying a complex figure or immediate recall of stories) are statistically removed from the memory scores, the residuals measure the amount of information retained or forgotten. Visual and verbal retention show opposite patterns of ability in TS girls and women, depending on whether they have the paternal or maternal X chromosome.[126] Compared to control females or males (who do not differ), visual retention is normal in those who have the

TABLE 9-5

DIRECT GENETIC INFLUENCES ON SEX DIFFERENCES IN BRAIN AND
BEHAVIORAL FUNCTION

Direct genetic effects on sex differentiation	Consequence	Evidence of effects on the brain
Genes unique to Y chromosome.	Expressed only in males.	Expression of Y-linked genes in two regions of human male brain. No expression in female brain.[1]
Recessive mutation of genes unique to the X chromosome (no homologous gene on the Y chromosome).	Recessive trait appears in all males who carry the mutation, but rarely in females (only if both X chromosomes carry the mutation).	Multiple recessive mutations on X associated with mental retardation in males.[2]
Genes unique to X that are active on both X chromosomes of females (one random X is inactivated in each female cell, but not completely).	Greater dosage of gene product in female than male cells.	In a region of the short arm of the X chromosome, genes expressed on both X chromosomes of human females are necessary for normal female brain anatomy, chemistry (greater female expression), and behavior.[3]
Different expressions of homologous genes on X and Y chromosomes.	Timing, dosage, or region of expression is not the same for XX and XY cells.	In mouse brain, expressions of homologous genes on X and Y are not expressed in parallel, and expression is greater in XX than in XY cells.[4]
Imprinted X-linked genes (expression is limited to a gene from one parent).	Paternal X genes only affect females. Maternal X genes affect both sexes, but are inactivated in half of female cells.	Girls and women with only a single X chromosome (X0) differ in behavioral functions depending on whether their single X comes from the father or mother.[5,6] Imprinted X genes have been identified in animals.[7]

SOURCES: 1. Mayer et al. 1998; 2. Rogers and Hamel 2005; 3. Good et al. 2003; 4. Xu and Disteche 2006; 5. Bishop et al. 2000; 6. Skuse et al. 1997; 7. Davies et al. 2006.

maternal X, but impaired in those who have the paternal X. In contrast, verbal retention is normal in those who have the paternal X, but impaired in those who have the maternal X.[127] Control females surpass control males in verbal retention, which is consistent with a cross-cultural female advantage in story recall.[128] The observations suggest that genes on the maternal X chromosome, which both sons and daughters receive, contribute to retention of complex visual information, whereas genes on the paternal X chromosome, which only daughters receive, contribute to retention of stories.

In summary, genetic factors contribute to sex differences in the brain and behavior both indirectly, through the influence of gonadal hormones, and by direct effects of genes on the sex chromosomes. Evidence from human and animal studies shows that nature actually utilizes all the various mechanisms available to differentiate males and females. However, although differences in sex roles in our evolutionary history may have put differing selective pressures on cognitive abilities in men and women, it is apparent that the sexes differ much less in these abilities than they do in play activities, interests, and goals. Species survival depends on effective reproductive and parenting behaviors, which necessarily entail differences between the sexes in brains and behavior. Evolution was faced with selecting characteristics that assured these differences but at the same time assured sufficient commonality in human cognitive abilities that either sex could survive, and even thrive, on a hunt or at the home base in the absence of the other. Briefly, selection for cognitive sex differences was probably constrained by selection for mental abilities that define our common humanity.

Individual Differences in Hormone Levels and Cognitive Abilities

Not all normal males have the same levels of androgens, nor do all females have the same levels of estrogen, progesterone, or testosterone. So in addition to variations in hormone levels across individuals before birth, there are lifetime variations from person to person in adulthood. These individual levels, while they do vary within a person from time to time, also show stability. The base level of testosterone, for example—the "rank" with respect to other individuals—remains fairly constant over both short- and long-term intervals.[129] This consistency appears to have a genetic component.[130]

Studies of the relation between individual differences in cognitive patterns of men and women and levels of sex hormones have mainly focused on testosterone (T). Females, of course, have much lower levels than males, with very little overlap between normal men and women. T levels can be measured from blood or from saliva. Salivary assays have become the preferred mode when studying normal volunteers because they are not invasive and pose no risk of infection.

The correlation between T levels and spatial ability is not linear.[131] That is, abilities do not simply increase, *pari passu*, with T levels. Rather, among normal men, those with higher T levels actually perform worse than those with lower levels (see figure 9-6). This fact is consistent with observations of CAH males. CAH males experience higher androgen levels prenatally than healthy males, but their spatial ability is not enhanced. In fact, some studies report that performance is poorer than in unaffected men.[132] Studies in rats also find that excess testosterone during early development diminishes rather than augments masculinization.[133]

FIGURE 9-6

RELATION OF SPACIAL ABILITY TO TESTOSTERONE (T)
LEVELS IN YOUNG MEN AND WOMEN

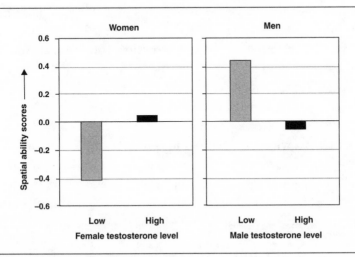

SOURCE: Adapted from Gouchie and Kimura 1991.
NOTES: The zero line represents the mean spatial score of the entire sample. High-T women (upper half of women) have better scores than low-T women, but low-T men (lower half of men) have higher scores than high-T men.

Among normal women, whose base level of T is low, those with higher T levels perform better on spatial tasks than those with lower levels (figure 9-6).[134] Again, this is consistent with the findings from CAH girls, in that pre-natal exposure to higher levels of T generally enhances spatial ability. It appears, therefore, that superior spatial ability is associated with an optimal level of T, neither too low nor too high, apparently in the lower range of normal Caucasian males. (We have no equivalent information for non-Caucasians.) Other abilities investigated to date appear not to be strongly associated with T levels. However, recall that male infants who make the least eye contact, and differ most in this trait from female infants, were exposed prenatally to T levels at the low end of the male range.[135] Those exposed to T levels at the high end of the male range were more similar to females in making more eye contact.

The cited studies, which divide each sex into high-T and low-T groups, consistently find that spatial ability is best when T is at an intermediate level. Future research may further refine the curvilinear relation between testosterone and spatial ability. The findings have been challenged on grounds that they would not be expected,[136] but no scientific basis nor even an argument is offered as to why a nonlinear relation should be unexpected. Indeed, given our knowledge that all brain chemicals (neurotransmitters) have an optimum level and that levels too low or too high disrupt brain function, it is unreasonable to expect that the higher the T level that bathes the brain, the better it functions in spatial cognition. A linear correlation between brain function and levels of neurotransmitters, drugs, or any other chemical does not characterize any known substance.

Fluctuations within Individuals in Hormones and Cognition

Hormone levels not only vary among individuals, but also within individuals over the course of daily, monthly, or yearly cycles. As we discuss below, these fluctuations influence cognition and behavior.

Estrogen. The role of estrogen in connection with differing abilities has been studied primarily by associating changes in estrogen across the menstrual cycle with changing cognitive patterns. We know that estrogen (higher in females than in males) is high in two phases of the cycle. First, in midcycle, just before ovulation, there is an estrogen "peak." This may last only one or two days, then it falls rapidly. This period is best identified by hormonal assay. Second, estrogen rises again, not as sharply, for a longer period about a week prior to next menstruation, falling again as the beginning of menses approaches (midluteal phase). The latter period is accompanied by a rise in progesterone (see figure 9-7). The midluteal phase can be moderately well identified from the actual onset of next menstruation although, again because of large variability, assays will improve accuracy. Either or both phases may be compared to the low-estrogen phase, which lasts for several days after onset of menstruation.

FIGURE 9-7

ESTROGEN AND PROGESTERONE LEVELS DURING
PHASES OF THE HUMAN MENSTRUAL CYCLE

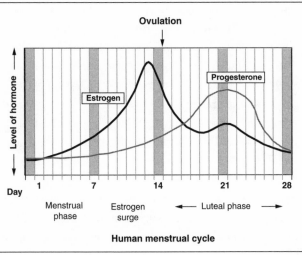

SOURCE: Adapted from Ferin et al. 1993.

Early investigations of cognitive changes across the cycle generally did not always use assays, and, moreover, used days *since* menstruation to identify phases. This method is unreliable, because women vary in the length of their menstrual cycles—twenty-eight days is merely an average—and also vary within themselves.[137] Early studies also employed behavioral tests that did not necessarily differentiate males from females and thus were less likely to be sensitive to effects of sex hormones. They emphasized concepts like attention, arousal, and automaticity.[138] In the past two decades, researchers have usually determined estrogen levels by actual measurement in blood or saliva, and have taken into account precise menstrual phases. Additionally, the choice of behavioral tests that differentiate males and females has yielded more reliable associations with changes in sex hormone levels.

When women's performance is compared between high-estrogen and low-estrogen periods, two findings are common:[139] First, the high-estrogen period is associated with enhancement of performance on certain tasks that favor females (finger dexterity, manual sequencing, oral fluency, and so forth) but not those that favor males.[140] Second (and conversely), the low-estrogen period is associated with enhancement of performance on certain spatial tasks that favor males but not on those that favor females.[141] Since the direction of the effect differs depending on what type of ability is compared, the results are not readily explained by general changes in mood or arousal level.

Changes in cognitive functions with variations in estrogen levels necessarily arise from systematic changes in the brain systems mediating these functions. Estrogen binds to receptors in brain cells and is transported to the cell nucleus, where it regulates gene expression. Depending on which genes are turned on or off, the chemistry and physiology of the cell and its interactions with other brain cells change. Which genes are turned on or off in which regions of the brain and how the brain changes are subjects of ongoing research.[142]

Testosterone. We have known for some time that testosterone levels vary across seasons of the year—they are higher in the autumn than in the spring, at least in the Northern Hemisphere, for which we have seasonal data.[143] This may be related to the advantage in preliterate times of conceiving in the autumn, so that births occurred in summer when food was more abundant. Concomitantly, the sperm count rises in the autumn.[144] Other studies have shown that T levels also change within men in a daily cycle[145] and even in a monthly cycle.[146]

Given that we find spatial ability higher in men with low-normal levels of T, we might expect a man's spatial performance to be lower in autumn (high T) than in spring (low T). Research indicates this is so. Young men tested in the spring, when their measured T levels are lower, perform better on spatial tests than those tested in the autumn. There is no seasonal influence on other tests, such as those that favor women or that generally show no sex difference.[147] Similar results on typical spatial tests are obtained for men and women across a lunar (monthly) cycle: Men perform better in their low-T phases and women in their high-T phases.[148] Unfortunately, no nonspatial results were reported in the latter study, so we cannot conclude that the effect is specific. Across a circadian (daily) cycle, men also perform better on spatial (but not nonspatial) tests at the time of day when T levels are low.[149]

These and other empirically established hormonal–cognitive relationships have oddly been described as "fairly murky,"[150] a critique that is difficult to cast in a scientific framework. In fact, the findings are quite robust and consistent. The evidence is that in normal young adults, individuals' abilities vary from time to time as hormone levels undergo natural fluctuations across daily, monthly, and annual rhythms. The specific cognitive changes depend on which hormone varies, whether the abilities are those that favor males or females, and probably other, as yet undiscovered, factors.

These fluctuations on a fairly stable baseline are assumed to relate to reproductive rhythms, and the cognitive changes may be only incidental. However, it is possible that some changes in cognitive patterns are themselves adaptive. For example, in early man's prehistory, having better spatial ability in spring may have been beneficial when setting out to hunt or to search for new campgrounds. Whatever the ultimate basis, the fact that such fluctuations affect cognitive patterns is a strong argument for the involvement of sex hormones in our cognitive makeup.

We must keep in mind that many of the associations we have described between hormone levels and cognitive performance are only correlational, which means that they occur together. To make a stronger *causal* connection— to attribute the cognitive changes to the hormonal changes—we need other supportive information. It is pertinent that the effects appear quite specific to certain abilities and not others, whether examining estrogen across the month or testosterone across seasons, months, or times of day. The changes aren't random, but are predictable from knowledge of the effects of sex hormones in general.

In addition, there are a number of studies in which hormones such as estrogen or testosterone are administered to individuals for a variety of therapeutic purposes. The administration of testosterone for a period of several weeks appears to improve spatial ability in subjects in whom the hormone was previously lower than optimal—for example, in older men[151] and in female-to-male transsexuals.[152] Again, the cognitive effects tend to be specific. So, while this kind of research is still sparse, such findings support the view that the cognitive changes accompanying natural hormonal fluctuations result from the hormones acting on brain systems.

Sex Differences in the Brain

Since all behavior is regulated by the nervous system, established and reliable sex differences in cognitive patterns are necessarily mediated by differences between the sexes in brain function. This is true no matter what the source of the differences, whether primarily experiential or primarily biogenic. Early research focused on anatomical brain differences, on the assumption that a sex difference for a particular skill would probably appear as a larger area of relevant cortex in the group advantaged for that particular ability. But this appears to be too simplistic a view, since some abilities—certain motor skills, for example—might be enhanced by being focally organized within a restricted area of the brain. In any case, sex differences need not manifest as a visible structural dichotomy in brain anatomy, though there are, in fact, many anatomical differences between men and women.

The most salient structural difference between the sexes is brain size, with men's brains being, on average, about a hundred cubic centimeters larger than women's. This appears not to be entirely accounted for by body size.[153] Within each sex, there is a small but consistent relation between brain size and intelligence.[154] Although there is a growing literature on possible sex differences in general intelligence, such differences, if they exist, are less than two or three IQ points, which is less than the average test–retest difference in individual scores.[155] Since the focus of this chapter is not on overall intelligence (g), we will leave aside brain size and focus on systems that might relate to the restricted cognitive sex differences we have described.

The neurological investigations include strictly anatomical (structural) studies, research on the asymmetric functions of the left and right brain hemispheres, and research on the commissures connecting the two hemispheres.

Studies on the cognitive effects of pathology (such as strokes and tumors) in particular parts of the brain have made a major contribution to understanding the way men's and women's brains may differ. More recently, brain imaging techniques, with which we can look at activation patterns in the brains of healthy people during the solution of various problems, have been employed to study sex differences in brain function. None of these methods has yet provided anything more than speculative answers to the question of what sex differences in brains underlie sex differences in cognition or behavior. We will, therefore, touch only briefly on brain sex differences.

Sex Differences in the Functional Asymmetry of the Two Hemispheres. For most right-handed people of either sex, the left cerebral hemisphere predominates in speech and related motor functions, and the right hemisphere in perceptual and spatial functions. It is often claimed that men have more asymmetric brain organization than women—that is, that women are more bilaterally organized. This would mean that in women, both hemispheres are more involved in language and/or visuospatial abilities than they are in men.

The origin of this idea is the fact that after strokes (which are reductions of blood supply to the brain) affecting the left hemisphere, a higher percentage of men than women have speech disturbances ("aphasia"). This led some researchers to conclude that speech must depend in part on the right hemisphere in women, protecting them from aphasia when the left is damaged. However, a strong argument against this conclusion is that women with damage to the right hemisphere rarely suffer aphasia, and no more often than do men. In one consecutive series of adult right-handers with right-hemisphere damage, only 1 out of 84 women and 2 out of 105 men suffered aphasia.[156]

Instead, it appears that men and women may depend for speech on different regions of the left hemisphere.[157] Women are more often rendered aphasic by anterior (front area) damage, and men by posterior (back area) damage (see figure 9-8). This may account for the higher incidence of aphasia in men, because when strokes result in restricted pathology rather than hemisphere-wide damage, it is more often the posterior region that is affected. Note that we are talking about very basic speech functions here—naming objects, counting, understanding simple sentences, and so on.

Not only speech functions but also related motor functions, such as the control of nonspeech oral and manual movements, depend differentially in

FIGURE 9-8

**PERCENTAGE OF MALE AND FEMALE PATIENTS WITH ANTERIOR OR POSTERIOR
LEFT-HEMISPHERE PATHOLOGY WHO SUFFER APHASIA OR APRAXIA**

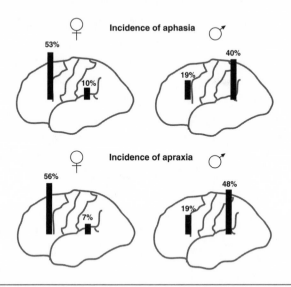

SOURCE: Kimura 2006.

women and men on anterior and posterior regions of the left hemisphere.
Thus, impaired ability to generate or reproduce certain hand and arm movements (called "apraxia") shows a similar pattern to aphasia (see figure 9-8).
The question is whether this difference in brain organization makes any difference in ability patterns in normal people. Some evidence suggests that it
does. We earlier described some motor skills that differentiated men and
women, with women excelling at intrapersonal skills and men at extrapersonal (see table 9-1). Men's demonstrably greater reliance on vision in control
of certain motor activities may be related to the fact that visual systems are
closer anatomically and functionally to the posterior region, which predominates in men, than to the anterior region, which predominates in women. In
contrast, the primary motor systems are closer anatomically and functionally
to the anterior region than to the posterior region.

As suggested by Levy and Heller,[158] when we look beyond basic speech
ability at tasks more verbal than motor, there is some evidence from a large

series of cases with damage to one hemisphere that some abilities are more bilaterally organized in women than in men. These include vocabulary, on which the scores of normal men and women do not differ,[159] and verbal fluency, on which women have the advantage.[160] Women's performance on these tasks is reduced by damage to either hemisphere, but men's only by damage to the left.

A similar pattern appears for verbal fluency after the left or right hemisphere is anesthetized by injection of sodium amytal into the carotid artery on one side.[161] Studies of a small group of presurgical patients reveal that men have a reduction in fluency from a previous baseline only when the left hemisphere is anesthetized, whereas women have lowered fluency when either side is anesthetized. No such sex difference appears for basic speech testing; aphasic errors in both sexes appear only after left-sided injection.

Verbal memory, on which normal women are advantaged, appears just as dependent on the left hemisphere in women as in men.[162] Similarly, a mental rotation task[163] and a line orientation task,[164] both advantaged in men, appear equivalently affected by damage to the right hemisphere in women and in men.

These combined results suggest that degree of functional brain asymmetry does not inevitably relate to sex differences in cognitive pattern, though it may do so for some particular cognitive functions.

The Cerebral Commissures. There has been considerable interest in whether the bridges of fibers that interconnect the two hemispheres of the brain (the cerebral commissures) are larger in one sex than the other. Each of two studies examined the small anterior commissure (AC) in large samples of 100 postmortem brains, but one reported a larger AC in women by 1.16 m^2 [165] and the other a larger AC in men by 1.07 m^2.[166] The average of the two studies suggests that if all brains in the population were measured, there would be no sex difference in the size of the AC.

Many studies have examined sex differences in the corpus callosum (CC), a massive commissure that contains more fibers than any other fiber tract in the brain. Although one group of researchers has reported that the back part of the corpus callosum, the splenium, is larger in women than men, this finding is not supported by many other investigations.[167] Claims made in the abstracts of studies that the splenium is larger in women than men often conflict with the full articles, which describe sex differences only in the ratio of the splenium to the size of the brain or total callosum, but not in the actual size of the splenium. As Driesen and Raz demonstrated in their large meta-analysis, the splenium is larger in men, but the ratio of the splenium to brain size is larger in women, which is a sex difference in the shape of the corpus callosum.[168] In any case, attempts to relate callosal size or shape to sex differences on cognitive tasks have so far been unsuccessful.[169]

Brain Imaging Studies. Different regions of the brain become active during different cognitive tasks, such as solving language or visuospatial problems. Functional brain imaging techniques, including functional magnetic resonance imaging (fMRI) and positron emission tomography (PET), among others, allow visualization of regional activations. Because the brain is constantly active, the change in activity introduced by any one cognitive task is not easy to detect. Consequently, a comparison condition is usually used which is intended to resemble the experimental condition in all but one cognitive process. The residual activation in the test condition after subtraction of activation in the control condition is assumed to identify the brain regions involved in the process of interest. It is likely, however, that the introduction of a new requirement does not merely add activation to a baseline task, but also changes activation dynamics. Additionally, as we discuss below, there are serious problems of interpretation even if the additivity assumption is valid.

A much-cited study on how men's and women's language functions are represented in the brain employed several conditions, all of which entailed comparing a pair of visual stimuli and pressing a button if they matched. Two of the tasks required making decisions about patterns of upper- and lower-case letters (assumed to involve visual and orthographic processing) and the rhyming of nonsense words (assumed to involve visual, orthographic, and phonological processing). Brain activation on the orthographic task was subtracted from brain activation on the rhyming task. The remaining activation was attributed to phonological processing. A significant sex difference was claimed in the pattern of phonological brain activation, namely bilateral activation in women and left-hemisphere activation in men.[170] This result is difficult to interpret because we do not know the pattern of activation on the orthographic task. Suppose both men and women rely on the left hemisphere on the rhyming task, since this has been demonstrated in patients with surgical disconnection of the two cerebral hemispheres.[171] If, on the orthographic task, men rely more on the right hemisphere and women on the left, which is not unlikely, then subtraction of the orthographic from the phonological task would artifactually leave bilateral activation in women and left-hemisphere activation in men.

Several studies have attempted to assess brain activation during mental rotation tasks, on which men on average typically outscore women. Although none of the tasks employed in these investigations revealed a significant sex difference, all had trends in favor of men. The insignificant sex differences are probably due to very small sample sizes, and perhaps to some quite difficult speeded conditions (which resulted in scores below chance in one study).[172] Although no consistent pattern of brain activation was found across all studies, the response to mental rotation after subtracting a baseline task generally showed bilateral activation in frontal and parietal areas of the brain, with slightly more parietal activity in men and slightly more frontal activity in women. The lack of a significant sex difference in mental rotation scores, however, limits the interpretation of these activation patterns as mechanisms for the cognitive differences. Obviously, since men and women differ in mental rotation ability, so, too, do the brain mechanisms that underlie the ability, but we have yet to identify them.

Men and Women: Values, Interests, and Talents

A report of the National Academy of Sciences makes the remarkable claim that "studies of brain structure and function, of hormonal modulation of performance, of human cognitive development, and of human evolution have not found any significant biological differences between men and women in performing science and mathematics that can account for the lower representation of women in academic faculty and scientific leadership positions in these fields."[173] As we have discussed, behavioral sex differences are found across cultures, generations, and age groups, including infants, are strongly affected by level of prenatal androgen exposure, and are influenced by characteristic and fluctuating levels of gonadal hormones in adults. They pertain not only to cognitive abilities, but even more so to interests, personality characteristics, and vocational goals.

Interests, Values, and Vocational Goals. No matter how brilliant children may be in mathematical talent, if they are more interested in acting and writing than in proving theorems, they will not become mathematicians. A few years ago, one of the six comedy actors on Main Stage at Second City, Chicago's most famous comedy club, was a young woman who was a math prodigy, which was evident from early childhood. She was also a prodigy in writing (she was writing a novel at the age of seven) and acting (for which she won a first-prize trophy in elementary school). It was impossible to measure her IQ at the age of six because she went off the top of the scale on both the verbal and performance sections of the Wechsler Intelligence Scale for Children. At age ten, she told one of us (JL), "Dad wants me to be a mathematician just because I'm good in math. But I'm not going be a mathematician. I'd hate to sit in an office all day and prove theorems. I'm going to be a comedy actor or comedy writer." She was a mathematics major on a full scholarship during her college years, but as soon as she graduated, she joined an acting troupe, then Second City, and is now a comedy writer for a national television network.

Evans and colleagues assessed the interests of very large numbers of eleventh-grade students in the United States, Taiwan, and Japan.[174] They reported that in all three nations, boys have a greater interest than girls in math, science, and sports, whereas girls have a greater interest than boys in music, art, and language. A study of 104 young women who had aspirations in the twelfth grade for a career in math or physical science (male-dominated occupations) found that by age twenty-five, 82 percent had changed their aspirations to gender-neutral or female-dominated occupations.[175] Significant predictors of maintaining an aspiration for an occupation in mathematics or physical science are the intrinsic values the women place on these subjects. Those who perceive lower intrinsic values of math and physical science are likely to switch aspirations. We have noted that girls and women with CAH, who were exposed to high levels of androgen during fetal life, are much more likely than their sisters (effect size about 2.0) to be interested in male-typical hobbies and to aspire to male-dominated occupations.[176]

Among gifted children in the fourth to sixth grades, three times as many girls as boys rate language arts or reading as the most interesting subject, whereas many more boys than girls rate science or math as the most interesting subject.[177] Gifted men and women from the longitudinal Study of Mathematically Precocious Youth (SMPY) also differ greatly in their interests and values.[178] As seen in figure 9-9, gifted women place little value on areas most strongly valued by gifted men (theoretical, political, economic), and men place little value on the two areas most strongly valued by gifted women (social, aesthetic). Values are important predictors of whether college degrees at age twenty-three or occupations at age thirty-three are in math or science, humanities, or other areas. Theoretical values are positively associated with math and science majors and occupations (and math SAT scores), while social and religious values are negatively associated. Aesthetic values are positively associated with humanities majors and occupations (and verbal SAT scores).

FIGURE 9-9

RELATIVE VALUATION BY GIFTED WOMEN AND MEN OF SIX VALUE DIMENSIONS

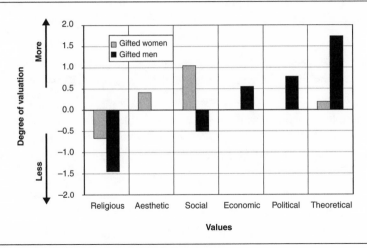

SOURCE: Adapted from Lubinski and Benbow 2006.

At age eighteen, SMPY high school seniors were asked to specify their intended undergraduate majors.[181] Among those who expected to major in a math or science area, the largest sex differences were in engineering and physical sciences, which were favored by boys, and biological and medical sciences, which were favored by girls. Thus, it is evident that even before they begin undergraduate studies, boys have a greater preference than girls for engineering and physical sciences, and girls have a greater preference than boys for biological and medical sciences.

A follow-up of these students when they were twenty-three years of age established that a majority actually majored and received undergraduate degrees in math and science areas. The same sex differences in preferences expressed in high school were manifested in undergraduate education. Of the math and science majors, men dominated in engineering and physical sciences and women in biological and medical sciences. At age thirty-three, nearly all these men and women were employed, and they expressed equally high job satisfaction and gave equally high ratings of their career success, although women devoted more time to families and less to work than did men.

In summary, interests and values differ between boys and girls and men and women and are important predictors of future college majors and occupations. From infancy to adulthood, females express and manifest more social interest than males, who manifest more interest in the physical world than females. Language arts and reading are more valued by girls, and math and science are more valued by boys. For students who choose careers in math and the sciences, males have a greater preference than females for math, the physical sciences, and engineering, whereas females have a greater preference than males for the biological, medical, and psychological sciences. These preferences of girls and women for the life and human sciences over math, the physical sciences, and engineering are consistent with an evolutionary history in which women were not only responsible for child care, teaching the young, and maintaining social cohesion, but were experts in the properties of plants for nutritional and medicinal purposes, and the deliverers of babies. The preferences of boys and men for mathematics and the physical sciences are consistent with their evolutionary role as hunters, weapon-makers, and explorers over large areas, which demanded interest in and attention to the properties of the physical world and the means to visualize and communicate them.

Talents and Achievements

The differences between the sexes in interests and values are correlated with differences in the levels and types of talents and achievements, which we discuss below.

Levels of Talent. Although the average advantage of men over women in mathematical reasoning is small (as shown in table 9-2), a small average advantage has major impacts at the upper end of abilities. The higher the ability level, the greater the disparity in the frequencies of men and women. This disparity is magnified further by the greater spread of male abilities.

Because academic careers are pursued by individuals with high levels of intellectual talent compared to the general population, we concentrate on the talents and achievements of the gifted participants in the SMPY. Children in the study were invited to take the SATs at age twelve or thirteen if they were in the upper 3 percent on standard achievement tests. Those selected for the SMPY scored at the upper 1 percent (Cohort 1), upper 0.5 percent (Cohort 2), or upper 0.01 percent (Cohort 3) of children their age. The higher the selection criterion, the greater the ratio of males to females.[182]

But does this matter for successful careers in mathematics and the sciences? Are people at the 95th percentile of mathematical talent just as likely to achieve high levels of success as those at or above the 99.99th percentile? Many have claimed there is no distinction in the achievements of people at a high level of aptitude and those with extremely exceptional talent above this.[183] The NAS report asserts that measures of mathematical aptitude such as the math SAT are not predictive of success in later science careers.[184]

Studies of the SMPY participants refute these claims.[185] The SAT scores obtained at age thirteen are strong predictors of adult achievement. At the twenty-year follow-up, when participants were thirty-three years old, Wai and colleagues selected the bottom and top quartiles of math SAT scores at age thirteen of those in the upper 1 percent of their sex (from Cohorts 1 and 2). Although even the lower quartile was in the top 1 percent at age thirteen, those in the upper quartile have more accomplishments at age thirty-three. Table 9-6 shows the differences between the bottom and top quartiles in doctoral degrees earned, income, patents held, and tenure at top-fifty universities. All the differences between these quartiles are reliable. They would occur by chance less than 1 percent of the time or, in the case of patents, less than 2 percent.[186]

A score at age thirteen that places a child in the upper 0.01 percent (1 in 10,000) of the population on either the math or verbal SAT or both (Cohort 3) predicts remarkable accomplishments in adulthood.[187] By their mid-thirties, 51.1 percent of these men and 54.3 percent of these women had earned doctoral degrees, which is more than fifty times the rate for the

TABLE 9-6
ACCOMPLISHMENTS OF SMPY PARTICIPANTS AT AGE 33

	Lower quartile of SMPY participants in the top 1 percent of SAT-M at age 13	Upper quartile of SMPY participants in the top 1 percent of SAT-M at age 13
Doctoral degrees	20%	32%
Income ≥ median for their sex	45%	56%
Hold patents	3.8%	7.5%
Tenure at a top-50 university	0.4%	3.2%

SOURCE: Wai et al. 2005.

general population. Over 50 percent of the doctoral degrees were earned at top-ten universities. Of twenty Master of Business Administration (MBA) degrees (5.3 percent), one was earned at a European university and all the rest at top-ten U.S. business schools. Approximately 60 percent of Cohort 3 had careers in medicine, law, science, engineering, or college teaching. Approximately 17 percent of the women and 9 percent of the men held tenured or tenure-track positions at universities. About half these positions were at top-fifty research universities, and of these, nearly 22 percent were already tenured. Patents were held by 17.8 percent of the men and 4.3 percent of the women. Of those whose top 0.01 percent placement was on the math SAT, 20.1 percent of men and 9.1 percent of women held patents. Men and women expressed comparably high satisfaction with their jobs, the direction of their careers, and perceived success in their careers.

A 2005 letter to *Science* signed by seventy-nine academic scientists and administrators claimed that "there is little evidence that those scoring at the very top of the range in standardized tests are likely to have more successful careers in the sciences. Too many other factors are involved."[188] To the contrary, the SMPY findings provide overwhelming evidence that SAT scores ranging from the upper 1 percent to 0.01 percent at age thirteen are strongly predictive of the children's subsequent successful careers. In doctoral degrees earned, tenure-track or tenured positions obtained, and patents held, the exceptionally gifted (top 1 in 10,000) greatly surpass those less gifted, and among the latter, those in the lowest quartile of the upper 1 percent have fewer achievements than those in the highest (see table 9-6).

As noted, the higher the extreme scores on the math SAT, the greater the ratio of males to females. More than eleven times as many boys as girls at age thirteen achieve math SAT scores of 700 or above (1 in 10,000 at this age). However, the girls who do score this high acquire as many doctorates as men and obtain tenured or tenure-track positions by their mid-thirties more frequently than men, perhaps because the men more often work as executives and administrators.[189] The ratio of male to female doctoral students in the inorganic sciences and engineering (see figure 9-10) reflects both differing interests of males and females and many fewer women at the extreme high end of mathematical talent and, as will be discussed, spatial ability. As the SMPY studies establish beyond question, the degree of talent among the gifted has a very large influence on adult achievements.

FIGURE 9-10

PERCENTAGE OF FEMALE DOCTORAL SCIENCE STUDENTS, 1976–2005

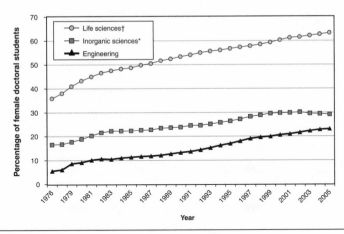

SOURCE: National Science Foundation/Resource Statistics.
NOTES: † = Biological and psychological sciences; * = Physical sciences, computer science, math. In engineering, the absolute number of women has been declining since 2003, but the decline in men is even greater.

The increase over the past thirty years in the percentage of science doctoral students who are women reflects the lowering of social barriers and the opening of opportunities. The changes in the life and human sciences, the inorganic sciences and math, and engineering are, however, much less than the differences among these fields in the percentages of women, which have remained remarkably stable since 1976 and may now be increasing. Men and women are drawn to careers where their interests and talents converge. The relationships depicted in figure 9-10 confirm the differing interests and values of males and females, observed even in early childhood, and differences in the skills in which they excel.

Patterns of Talent. Scores on cognitive assessments across diverse areas are positively correlated, which reflects variations among people in a general factor of intelligence. The correlations are far from perfect, however, because individuals vary in special abilities. Park and his colleagues examined the relationships among adult achievements, ability level on the SATs at age thirteen, and ability tilt toward higher math or verbal SAT scores in SMPY participants. Those who obtained undergraduate or graduate degrees in the humanities or had faculty appointments in the humanities or creative literary accomplishments had a verbal tilt in SATs at age thirteen. Those with degrees or faculty appointments in math or science or who held patents had a math tilt.[190]

Males had a greater mathematical tilt than females (effect size $d = 0.72$, see figure 9-3), as well as a higher ability level (effect size $d = 0.40$). Because the average and variance of math ability are greater in males, they not only occur more frequently than females in the top 1 percent of the population, but their talent within this top 1 percent is higher. In each of the four areas of accomplishment, from four-year degrees to faculty positions and creative accomplishments, there are more gifted men than women in math and science areas and more gifted women than men in humanities. There are also more gifted women than men with MD degrees.

Among SMPY participants in the upper 0.01 percent, the number of accomplishments and awards in math and science compared to those in humanities and arts vary as a function of the relation of math and verbal SAT scores at age thirteen.[191] When the two scores are similar, an equal number of people are characterized by accomplishments or awards in math or science and arts or humanities. A predominance of accomplishments and awards in math and science or in arts and humanities is predicted, respectively, by higher math than verbal or higher verbal than math ability.

The verbal and math SATs assess two major cognitive dimensions, but neither the math nor verbal SAT directly assesses spatial reasoning. High levels of spatial ability are likely to characterize engineers, architects, experimental physicists, mathematicians who specialize in geometry or topology, organic chemists, crystallographers, and others who need excellent visualization skills. A few hundred SMPY children who were in the upper 0.5 percent in general ability were tested on the mechanical and spatial reasoning tests of the Differential Aptitude Test battery in the spring of their seventh-grade years. Children performed in the upper 5 percent of eighth graders, which is the earliest grade for which norms are available. Because spatial and mechanical reasoning are highly correlated, scores were combined into a single measure to examine the influence of spatial-mechanical ability on preferences and achievements.[192]

The favorite high school courses of the majority of boys were in math or science and the least favorite in the humanities and social sciences, preferences that were associated with higher math and spatial-mechanical abilities and lower verbal abilities. The favorite high school courses of the majority of girls were in the humanities and social sciences and the least favorite in math or science, preferences that were associated with reduced math ability, increased verbal ability, and very reduced spatial-mechanical ability.

Undergraduate and graduate degrees and occupational groups are predicted by ability patterns manifested at age thirteen. Degrees and occupations in engineering are predicted by relatively lower verbal ability, higher math ability, and exceptionally high spatial-mechanical ability. Those in physics, however, are predicted by increased math and spatial-mechanical ability and very high verbal abilities. Degrees and occupations in the humanities and social sciences are predicted by decreased math and spatial-mechanical abilities and increased verbal abilities. The strongest predictor of engineering is lower verbal ability and very high spatial-mechanical ability. To a lesser degree, the same is true of math and computer science (combined in the analysis), whereas physical sciences are associated with high verbal as well as high spatial-mechanical ability.

In the foregoing analyses, Shea and colleagues provided no information regarding men and women. However, spatial-mechanical reasoning differentiates men and women more than any other cognitive ability. It is, therefore, not surprising that graduate engineering is dominated by men to a significantly greater extent than is inorganic science (see figure 9-10).[193]

Facts and Fictions about Women in Science

The report of the National Academy of Sciences attributes the male dominance in academic mathematics, science, and engineering to unintended or unconscious biases against women, supposedly outdated institutional structures that demand total commitment to an academic career, and evaluative processes that give an unfair weighting to qualities that are supposedly stereotypically male. The "stereotypically male" qualities are said to be competitiveness, which is supposedly manifested in a devotion to research, and greater time spent in research than in the "nurturing activities" of teaching and mentoring.[194] We briefly address these attributions.

Unconscious Bias against Women? There are serious ethical issues in accusing people of unconscious bias, which, on the one hand, assumes guilt unless innocence is proved, and, on the other hand, denies the possibility of such proof.[195] A frequently cited paper that purportedly establishes biases against women in the awarding of graduate fellowships by the Swedish Research Council employs illegitimate statistical procedures and fails to establish what it claims (see appendix).[196] Moreover, a study of fellowship awards in 1998 by the European Molecular Biology Organization (EMBO) found no evidence of systematic bias against women.[197] Among those awarded fellowships, men surpassed women in the impact of their publications and in the number of first-authored publications, whereas women surpassed men in the number of total publications. Among applicants, men surpassed women in all three factors (impact of publications, first-authored publications, and total publications).

Unreasonable Demands of an Academic Career? Although success in any demanding career necessarily requires time, energy, effort, and commitment, academic institutions are far more lenient than most businesses in giving time off and leaves of absence. The assertion that academic institutions require "total" commitment to an academic career, as if no time were allowed for anything else other than sleeping, eating, and personal care, is simply false.

Moreover, unless the authors of the NAS report offer evidence that academic careers in mathematics, the inorganic sciences, and engineering demand far more commitment than do those in the life and human sciences, it is illegitimate to ascribe the lower representation of women to unreasonable requirements for commitment. As barriers against them have relaxed over the last quarter-century, women have come to dominate or are coming to dominate the social sciences, psychology, and biomedical sciences (see figure 9-10).

Perhaps the authors of the NAS report would argue that the inorganic sciences and engineering fields unfairly discriminate against women to a greater degree than do the life sciences. First, women have increased their representation among veterinary students from 5 percent in the 1960s (compared to 9 percent of medical students) to nearly 80 percent today. If the freedoms, opportunities, and power that women have achieved during that time have allowed them to dominate a field once renowned for its bias against them, they would surely overcome much lesser bias in the inorganic sciences and engineering if these were areas that appealed to them. Second, larger gains occurred in the percentage of female doctoral students in the inorganic sciences and engineering in the eight-year period from 1990 through 1997, despite biases against women, than in the eight-year period from 1998 through 2005, when biases were reduced (see figure 9-10). This suggests that it is not bias against women that accounts for their low percentages in these fields, but rather the preference of the majority of women for the life and human sciences, which showed the most rapid increase in female percentage from 1976 to 1982 and a smaller but steady increase ever since. As Lubinski and Benbow have noted, we should not consider that math-talented women who do not choose to pursue mathematics, the inorganic sciences, or engineering are lost to these fields, but rather that they enrich the fields they do enter.[198]

Overvaluing of Research? The claim of the NAS report that a devotion to research reflects "stereotypically male" competitiveness, whereas teaching and mentoring are "nurturing activities" that are more characteristic of women and should be given greater weight in academic evaluations, is not only denigrating to female scientists, but a threat to scientific achievement by both women and men. There are no world-class scientists or mathematicians who are not passionate about the research endeavor. Teachers of science and mathematics who do not advance their fields through research are teachers and not scientists or mathematicians. Mentors of doctoral and postdoctoral science and mathematics students cannot train these students in research unless they themselves are active researchers. Academic genealogies testify to the critical importance of a mentor's success as a researcher to the success of the students trained. For example, the 2004 Nobel Prize in Biomedical Sciences was awarded to Richard Axel and Linda Buck for their work on the neurobiology and genetics of olfaction. Axel received an MD and then was trained in research by Sol Spiegelman, a renowned researcher who received the Lasker Award (one step from a Nobel). Buck's doctoral advisor was Ellen Vitetta, another renowned and highly active researcher who is a member of the National Academy of Sciences. Mentors of great scientists are almost always internationally recognized for outstanding research.

Summary. The fundamental claim of the NAS report that the greater frequencies of men than women in engineering, the physical sciences, and mathematics are due entirely or predominantly to social barriers against women has no scientific foundation. There is no evidence that speeded association tests[199] measure "unconscious bias" rather than experience. To make a faster association between "engineer" and a man's photograph than a woman's photograph does not establish bias, unconscious or otherwise. Moreover, there is no evidence that such speeded associations have any relation to conscious decision making. There is no evidence that the academic demands of engineering, the physical sciences, and mathematics place a greater burden on women than do the medical, veterinary, life, or human sciences. The idea that a great new generation of scientists could be trained by mentors who place more value on nurturing activities than on research and therefore do little or no meaningful research themselves is absurd on its face, inconsistent with academic histories, an insult to female scientists, and a serious danger to science itself.

Conclusions

We have reviewed overwhelming evidence that genetic and hormonal differences between males and females are major causes of sex differences in behavior. These include differences in social behaviors in infants, play behaviors, interests, activities, educational and vocational goals, choices of occupations, patterns of cognitive abilities, and the frequency of extreme giftedness in spatial, mechanical, and mathematical ability. The dominance of female doctoral students in the life and human sciences and of male doctoral students in the inorganic sciences and engineering is consistent with and predictable by sex differences in interests and ability patterns. The greater social interest and ability of females than males is evident in infancy. The differing play activities and interests of boys and girls share similarities with sex differences in the play behaviors of nonhuman primates. Interests, activities, values, and vocational goals that differentiate girls from boys and women from men are strongly affected by the level of fetal androgen exposure or tissue sensitivity to androgen. Daily, monthly, or yearly cycles in levels of adult sex hormones influence performance on certain verbal and spatial tasks

Although the magnitude of average sex differences in certain cognitive abilities has declined in the last forty years, none of these differences has disappeared or is likely to disappear. However, even if there were no cognitive sex differences in average mathematical or spatial ability, there would still be more males than females at the upper end of intellectual talent due to greater male variance. In consequence, there would still be more males than females who meet even minimum standards to be academic engineers, physical scientists, or mathematicians, and many more men than women with exceptionally high levels of talent. Even if both the mean and variance of mathematical and spatial ability were identical for men and women, there would be more men than women with interests in engineering, the physical sciences, or mathematics, and more women than men with interests in medicine, the life and human sciences, humanities, or the arts. Because interests, values, and vocational aspirations are strongly influenced by the organizing actions of androgen on the fetal brain, we could not—without coercive force or manipulation of the hormonal environment of the fetus—equalize the numbers of men and women in all fields of science, engineering, and math even if, in contrast to reality, men and women were identical in the mean and variance of all cognitive abilities.

Appendix to Chapter 9

To determine whether the sex of the applicant plays a role in the awarding of fellowships, all objective evaluative criteria (for example, number of publications, number of first-author publications, quality of journals in which work is published, impact of the publications, and so on) must be entered into a multiple prediction equation, along with the sex of the applicant (coded numerically as 0 or 1 or any other pair of numbers). The equation predicts success in winning a fellowship and establishes the weight of each predictor. The question is whether the accuracy of predicting the award of a fellowship is improved when the applicant's sex is added as a predictor. In a study published in 1997 in which they claimed to have found bias in the awarding of fellowships to female scientists in Sweden, Wennerås and Wold failed to apply this analysis. Instead, they combined the applicant's sex with a single objective criterion in a series of separate predictions. This is illegitimate for two reasons. First, omitted objective criteria in any given equation may be differentially associated with one or the other sex, in which case causation is spuriously attributed to the sex of the applicant. Second, because the different objective criteria are correlated with each other, it is illegitimate to conduct separate statistical analyses as if each criterion were independent of others.

When John Steiger, a statistician at Vanderbilt University, criticized the failure of Wennerås and Wold to employ multiple regression analyses, Wold, in an interview with reporter John Tierney of the *New York Times* (Summer 2008), said, "This is a very puzzling remark. As anyone using multiple regression knows, the X variables entered into the equation are supposed to be independent of one another (hence, the name 'independent variables')." This statement reflects a complete misunderstanding of multiple regression and of the meaning of "independent variable." Interactive contributions arise to the extent that predictor variables are correlated. Indeed, one of the major strengths of multiple regression is that it separates the independent from the interactive contributions of predictor variables. Steiger, in repeated letters of request during 2007, sought access to the data so he could analyze it appropriately, but never heard from Wennerås and was finally informed by Wold that the data were lost. Wold told Tierney in 2008, however, that "Ulf Sandström recently did a reanalysis of our data using more 'novel' measures of productivity." [200]

Notes

1. Halpern et al. 2007; Ceci and Williams 2006; National Academy of Sciences et al. 2007.
2. Domenice et al. 2003.
3. Byskov 1986; McLaren 1991; Matzuk et al. 2002; Ottolenghi et al. 2007.
4. Gartler et al. 1972; Gartler et al. 1975.
5. Simpson and Rajkovic 1999; Toniolo and Rizzolio 2007.
6. Reynaud et al. 2004.
7. Imperato-McGinley and Zhu 2002.
8. Creighton 2004.
9. See chapter 7, above.
10. Daly and Wilson 1983, table 10-1; Lee and DeVore 1968.
11. See Daly and Wilson 1983, table 10-1 for a more detailed outline.
12. Blumenschine and Cavallo 1992; Lovejoy 1981.
13. Ecuyer-Dab and Robert 2004.
14. Brandner 2007.
15. A statistically significant difference is one that is very unlikely to be due to chance. We can then infer that a difference observed in the sample is real and can be generalized to the population. Descriptions of findings refer to population inferences unless otherwise stated.
16. College Board 2007.
17. Machin and Pekkarinen 2008.
18. Kail and Siegel 1977; Postma et al. 1999; Sandström and Lundberg 1956.
19. Saucier et al. 2007.
20. Voyer et al. 2007.
21. Hampson et al. 2006; Rosenthal et al. 1979; Proverbio et al. 2007.
22. LaFreniere and Dumas 1996; Bosack and Astington 1999.
23. Murphy and Ross 1987; Bosack and Astington 1999.
24. Lindeman 1991.
25. Ciarrochi et al. 2005.
26. Podrouzek and Furrow 1988.
27. Hall 1984.
28. Schirmer et al. 2005.
29. Ford 1982.
30. Jobson and Watson 1984.
31. Engelhard 1990.
32. Kimura 1994; Lynn and Irwing 2002.
33. Low and Over 1993; Lummis and Stevenson 1990.
34. Felson and Trudeau 1991.
35. Wainer and Steinberg 1992.
36. Halpern et al. 2007; Valian 2006.

37. College Board 2007.

38. Lewis and Willingham 1995.

39. Feingold 1992; Hedges and Nowell 1995; College Board 2007.

40. Out of a possible score of 800.

41. Lubinski and Benbow 2006.

42. International Mathematical Olympiad 2008.

43. Mathematical Association of America 2008.

44. Law et al. 1993.

45. Sanders et al. 2007.

46. Watson and Kimura 1991.

47. Chipman et al. 2002.

48. Lynn 1994.

49. Bradshaw and Nichols 2004; Johnsson-Smaragdi and Jönsson 2006.

50. Kobrin and Kimmel 2006.

51. Implicit memories refer to mental representations of past experiences that enhance task performance but are not consciously remembered.

52. Vandenberg and Kuse 1978.

53. Halpern et al. 2007.

54. Collins and Kimura 1997.

55. Galea and Kimura 1993; Moffat et al. 1998.

56. Saucier et al. 2002.

57. Robert and Harel 1996; Robert and Ohlmann 1994; Thompson et al. 1981; Wittig and Allen 1984.

58. Thompson et al. 1981; Wittig and Allen 1984.

59. Newcombe 2006.

60. Bennett et al. 1961; Feingold 1988. The DAT battery comprises tests of verbal reasoning, numerical ability, abstract reasoning, mechanical reasoning, space relations, spelling, language (grammar), and clerical speed and accuracy.

61. Feingold 1988, 100.

62. Nowell and Hedges 1998.

63. Feingold 1996.

64. Nowell and Hedges 1998.

65. Feingold 1996.

66. Vandenberg and Kuse 1978.

67. Masters and Sanders 1993.

68. Chipman and Kimura 1998; Geary and DeSoto 2001; Peters et al. 2006.

69. Baenninger and Newcombe 1989; Devlin 2004.

70. Munroe and Munroe 1971.

71. Baenninger and Newcombe 1989.

72. Terlecki 1995.

73. Mann et al. 1990.

74. Owen and Lynn 1993.

75. Chang and Antes 1987.
76. Mayes et al. 1988; Owen and Lynn 1993.
77. DeLisi et al. 1989.
78. Silverman and Phillips 1993.
79. Geary and DeSoto 2001.
80. Peters et al. 2006.
81. Pontius 1997a and 1997b.
82. Witkin 1950.
83. Berry 1966; MacArthur 1967.
84. Ekstrom et al. 1976.
85. Kimura 1994; Watson and Kimura 1991.
86. Moore and Smith 1987; Jensen 1988.
87. Engelhard 1990.
88. Lummis and Stevenson 1990.
89. Campbell 1991.
90. Connellan et al. 2001.
91. Gamé et al. 2003.
92. Lutchmaya and Baron-Cohen 2002.
93. Olafsen et al. 2006.
94. Gunnar and Donahue 1980.
95. Hittelman and Dickes 1979.
96. Lutchmaya et al. 2002.
97. Ibid.
98. Gouchie and Kimura1991; Moffat and Hampson1996; Ostatníková et al. 2002; Shute et al. 1983.
99. Diamond and Sigmundson 1997; Colapinto 2006.
100. Colapinto 2004.
101. DiPietro 1981; Whiting and Edwards 1973.
102. Goy et al. 1988; Goy and Resko 1972.
103. Feldman et al. 1977; Maestripieri and Pelka 2002.
104. Herman et al. 2003.
105. Alexander and Hines 2002; Hassett et al. 2004.
106. Goy et al. 1988; Goy and Resko 1972; Goy and McEwen 1980; Isgor and Sengelaub 1998; Isgor and Sengelaub 2003; MacLusky and Naftolian 1981; Roof 1993; Wallen 2005; Williams et al. 1990.
107. Clark and Goldman-Rakic 1989.
108. Bachevalier et al. 1989; Bachevalier and Hagger 1991; Hagger et al. 1987.
109. Gouzoules and Gouzoules 1989.
110. Wallen 2005.
111. Diamond and Sigmundson 1997; Colapinto 2006.
112. Reiner and Gearhart 2004.
113. Berenbaum and Sandberg 2004.

114. Hines, Ahmed, and Hughes 2003; Melo et al. 2003; Jürgensen et al. 2007; Wisniewski et al. 2000.

115. Udry et al. 1995; Ehrhardt et al. 1977; Hines et al. 2002.

116. Berenbaum 1999; Berenbaum and Bryk 2008; Berenbaum and Resnick 2006; Berenbaum and Snyder 1995; Cohen-Bendahan et al. 2005; Hines 2004; Hines et al. 2004; Iijima et al. 2001; Nordenstrom et al. 2002; Pasterski et al. 2007.

117. Cohen-Bendahan et al. 2005.

118. Hines, Fane, et al. 2003.

119. Berenbaum et al. 2006; Halpern et al. 2007; Mueller et al. 2008; Resnick et al. 1986.

120. Hines, Fane, et al. 2003; Baker and Ehrhardt 1974; Helleday et al. 1994; Malouf et al. 2006.

121. Puts et al. 2008.

122. Arnold 2004; Arnold and Burgoyne 2004.

123. Vawter et al. 2004.

124. Creswell and Skuse 1999; Marco and Skuse 2006.

125. Skuse et al. 1997.

126. Bishop et al. 2000.

127. Ibid.

128. Mann et al. 1990.

129. Dabbs 1990; Smals et al. 1976.

130. Meikle et al. 1987.

131. Shute et al. 1983; Gouchie and Kimura 1991; Moffat and Hampson 1996; Celec et al. 2002.

132. Hampson et al. 1998; Hines, Fane, et al. 2003; Berenbaum et al. 2006.

133. Goy and McEwen 1980.

134. Shute et al. 1983; Gouchie and Kimura 1991; Moffat and Hampson 1996.

135. Lutchmaya et al. 2002.

136. Halpern et al. 2007; Newcombe 2006.

137. Hampson and Young 2008.

138. Broverman et al. 1981; Komnenich et al. 1978. Automaticity refers to skilled performance in the absence of conscious attention.

139. Hampson 1990.

140. Saucier and Kimura 1998.

141. Hausmann et al. 2000; Silverman and Phillips 1993.

142. When these findings were first reported at a meeting in the late 1980s, the result was a media frenzy and generally negative commentary by female journalists, who felt that women were being singled out by this research and believed that the data would be used to bolster prejudices about women's instability and bring about policies detrimental to them. In fact, research on analogous hormonal fluctuations in men was already underway. We note, also, that scientific findings rarely influence political views. Rather, politics misuses science or creates a compatible pseudoscience

in service of its aims. In Hitler's Germany, Albert Einstein's papers were burned and "Aryan physics" was created. During China's Great Cultural Revolution, respected geneticists were yanked from their laboratories, placed in the cotton fields, and told to paint the cotton bolls blue so that each generation they would grow a little bluer and obviate the necessity to dye the cloth. And in North America today, the overwhelming scientific evidence that cognitive and behavioral sex differences partly reflect biological influences appears to have no effect on those who seek government policies aimed at parity of men and women in all educational programs and all fields of endeavor.

143. Smals et al. 1976; Dabbs 1990; Kimura and Hampson 1994; Ostatníková, Putz, and Matûjka 1995; Ostatniková, Putz, Matûjka, and Országh 1995.

144. Levine 1991.

145. Reinberg and Lagoguey 1978; Moffat and Hampson 1996.

146. Celec et al. 2002.

147. Kimura and Hampson 1994.

148. Celec et al. 2002.

149. Moffat and Hampson 1996.

150. Newcombe 2006.

151. Cherrier et al. 2000; Janowsky et al. 1994.

152. Slabbekoorn et al. 1999; Van Goozen et al. 2002.

153. Rushton and Ankney 2007.

154. Willerman et al. 1991; Andreasen et al. 1993.

155. Lynn 1994; Jackson and Rushton 2006; Deary 2007.

156. Kimura 1987.

157. Kimura 1983; Cappa and Vignolo 1988; Hier et al. 1994; Mateer et al. 1982; Buckner 1995.

158. Levy and Heller 1992.

159. Kimura and Harshman 1984.

160. Kimura 1999.

161. McGlone and Fox 1982.

162. Berenbaum et al. 1997; Kimura 1999.

163. Kimura 1999.

164. Desmond et al. 1994.

165. Allen and Gorski 1991.

166. Lasco et al. 2002.

167. Bishop and Wahlsten 1997; Driesen and Raz 1995.

168. Driesen and Raz 1995.

169. Davatzikos and Resnick 1998; Salat et al. 1997.

170. Shaywitz et al. 1995.

171. Levy and Trevarthen 1977.

172. Hugdahl et al. 2006; Jordan et al. 2002.

173. National Academy of Sciences et al. 2007.

174. Evans et al. 2002.
175. Frome et al. 2006.
176. Berenbaum and Bryk 2008.
177. Olszewski-Kubilius and Turner 2002.
178. Lubinski and Benbow 2006.
179. Achter et al. 1999.
180. Wai et al. 2005.
181. Webb et al. 2002.
182. Lubinski and Benbow 2006.
183. Muller 2005 et al.; Vasquez and Jones 2006.
184. National Academy of Sciences et al. 2007. Number 2-5 in listing of findings, 25.
185. Lubinski and Benbow 2006.
186. Wai et al. 2005.
187. Lubinski et al. 2006.
188. Muller et al. 2005.
189. Lubinski and Benbow 2006.
190. Park et al. 2007.
191. Lubinski and Benbow 2006.
192. Shea et al. 2001.
193. Ibid.
194. National Academy of Sciences et al. 2007.
195. Mitchell and Tetlock 2006.
196. Wennerås and Wold 1997.
197. Gannon et al. 2001.
198. Lubinski and Benbow 2006.
199. Carney et al. 2007.
200. Email from Wold to Tierney, July 21, 2008.

References

Achter, John A., David Lubinski, Camilla P. Benbow, and Hossain Eftekhari-Sanjani. 1999. Assessing Vocational Preferences among Gifted Adolescents Adds Incremental Validity to Abilities: A Discriminant Analysis of Educational Outcomes over a Ten-Year Interval. *Journal of Educational Psychology* 91: 777–86.

Alexander, Gerianne, and Melissa Hines. 2002. Sex Differences in Response to Children's Toys in Nonhuman Primates (*Cercopithecus aethiops sabaeus*). *Evolution and Human Behavior* 23: 467–79.

Allen, Laura S., and Roger A. Gorski. 1991. Sexual Dimorphism of the Anterior Commissure and Massa Intermedia of the Human Brain. *Journal of Comparative Neurology* 312 (1): 97–104.

Andreasen, Nancy A., Michael Flaum, Victor Swayze, D. S. O'Leary, R. Alliger, G. Cohen, J. Ehrhardt, and W. T. C. Yuh. 1993. Intelligence and Brain Structure in Normal Individuals. *American Journal of Psychiatry* 150: 130–34.

Arnold, Arthur P. 2004. Sex Chromosomes and Brain Gender. *Nature Reviews Neuroscience* 5: 1–8.

Arnold, Arthur P., and Paul S. Burgoyne. 2004. Are XX and XY Brain Cells Intrinsically Different? *Trends in Endocrinology and Metabolism* 15 (1): 6–11.

Bachevalier, Jocelyne, and Corinne Hagger. 1991. Sex Differences in the Development of Learning Abilities in Primates. *Psychoneuroendocrinology* 16 (1–3): 177–88.

Bachevalier, Jocelyne, Corinne Hagger, and Barry B. Bercu. 1989. Gender Differences in Visual Habit Formation in 3-Month-Old Rhesus Monkeys. *Developmental Psychobiology* 22: 585–99.

Baenninger, Maryann, and Nora Newcombe. 1989. The Role of Experience in Spatial Test Performance. *Sex Roles* 20: 327–44.

Baker, S. W., and A. A. Ehrhardt. 1974. Prenatal Androgen, Intelligence and Cognitive Sex Differences. In *Sex Differences in Behavior*, ed. R.C. Friedman, R. M. Richart, and R. L. Vande Wiele, 53–76. New York: Wiley.

Beatty, William W. 2002. Sex Difference in Geographical Knowledge: Driving Experience Is Not Essential. *Journal of the International Neuropsychological Society* 8 (6): 804–10.

Beatty, William W., and Alexander I. Troster. 1987. Gender Differences in Geographical Knowledge. *Sex Roles* 16: 565–90.

Benbow, Camilla P. 1988. Sex Differences in Mathematical Reasoning Ability in Intellectually Preadolescents: Their Nature, Effects, and Possible Causes. *Behavioral and Brain Sciences* 11: 169–82.

Bennett, George K., Harold G. Seashore, and Alexander G. Wesman. 1961. *Differential Aptitude Tests* (DAT). New York: Psychological Corporation.

Berenbaum, Sheri A. 1999. Effects of Early Androgens on Sex-Typed Activities and Interests in Adolescents with Congenital Adrenal Hyperplasia. *Hormones and Behavior* 35: 102–10.

Berenbaum, Sheri A., L. Baxter, M. Seidenberg, and B. Hermann. 1997. Role of the Hippocampus in Sex Differences in Verbal Memory: Memory Outcome Following Left Anterior Temporal Lobectomy. *Neuropsychology* 11: 585–91.

Berenbaum, Sheri A., and K. K. Bryk. 2008. Biological Contributors to Gendered Occupational Outcome: Prenatal Androgen Effects on Predictors of Outcome. In *Gender and Occupational Outcomes: Longitudinal Assessment of Individual, Social, and Cultural Influences*, ed. H. M. G. Watt and J. S. Eccles, 235–64. Washington, D.C.: APA Books.

Berenbaum, Sheri A., B. Fesi, and K. Bryk. 2006. Early Androgen Effects on Spatial and Mechanical Abilities. Poster presented at the Tenth Annual Meeting of the Society for Behavioral Neuroendocrinology, Pittsburgh, June.

Berenbaum, Sheri A., and Susan Resnick. 2006. The Seeds of Career Choices: Prenatal Sex Hormone Effects on Psychological Sex Differences. In *Why Aren't More Women in Science?* ed. S. J. Ceci and W. Williams. Washington, D.C.: APA Books.

Berenbaum, Sheri A., and D. E. Sandberg. 2004. Letter Regarding Reiner and Gearhart, 2004. *New England Journal of Medicine* 350 (21): 2204-6.

Berenbaum, Sheri A., and Elizabeth Snyder. 1995. Early Hormonal Influences on Childhood Sex-Typed Activity and Playmate Preferences: Implications for the Development of Sexual Orientation. *Developmental Psychology* 31: 31–42.

Berry, John W. 1966. Temne and Eskimo Perceptual Skills. *International Journal of Psychology* 1: 207–29.

Bieri, James, Wendy M. Bradburn, and M. David Galinsky. 1958. Sex Differences in Perceptual Behavior. *Journal of Personality* 26: 1–12.

Bishop, D. V. M., E. Canning, K. Elgar, E. Morris, P. A. Jacobs, and D. H. Skuse. 2000. Distinctive Patterns of Memory Function in Subgroups of Females with Turner Syndrome: Evidence for Imprinted Loci on the X-Chromosome Affecting Neurodevelopment. *Neuropsychologia* 38: 712–21.

Bishop, Katherine M., and Douglas Wahlsten. 1997. Sex Differences in the Human Corpus Callosum: Myth or Reality? *Neuroscience and Biobehavioral Reviews* 21: 581–601.

Blumenschine, Robert J., and A. J. Cavallo. 1992. Scavenging and Human Evolution. *Scientific American* 267: 90–96.

Bosack, Sandra, and Janet Wilde Astington. 1999. Theory of Mind in Preadolescence: Relations Between Social Understanding and Social Competence. *Social Development* 8 (2): 237–55.

Bradshaw, Tom, and Bonnie Nichols. 2004. Reading at Risk: A Survey of Literary Reading in America. Research Division Report No. 46. National Endowment for the Arts. June. http://arts.endow.gov/research/ReadingAtRisk.pdf (accessed January 23, 2009).

Brandner, Catherine. 2007. Strategy Selection during Exploratory Behavior: Sex Differences. *Judgment and Decision Making* 2 (5): 326–32.

Broverman, D. M., W. Vogel, E. L. Klaiber, D. Majcher, D. Shea, and V. Paul. 1981. Changes in Cognitive Task Performance across the Menstrual Cycle. *Journal of Comparative and Physiological Psychology* 95: 646–54.

Buckner, R. L., M. E. Raichle, and S. E. Petersen. 1995. Dissociation of Human Prefrontal Cortical Areas across Different Speech Production Tasks and Gender Groups. *Journal of Neurophysiology* 74: 2163–73.

Byskov, Anne Grete. 1986. Differentiation of the Mammalian Embryonic Gonad. *Physiological Reviews* 66 (1): 71–117.

Campbell, James Reed. 1991. The Roots of Gender Inequity in Technical Areas. *Journal of Research in Science Teaching* 28: 251–64.

Cappa, S. F., and L. A. Vignolo. 1988. Sex Differences in the Site of Brain Lesions Underlying Global Aphasia. *Aphasiology* 2: 259–64.

Carney, D. R., B. A. Nosek, A.G. Greenwald, and M. R. Banaji. 2007. Implicit Association Test (IAT). In *Encyclopedia of Social Psychology*. ed. R. Baumeister and K. Vohs. Thousand Oaks, Calif.: Sage. Test available online at https://implicit.harvard.edu./implicit/research/ (accessed May 4, 2009).

Ceci, Stephen J., and Wendy M. Williams, eds. 2006. *Why Aren't More Women in Science?* Washington, D.C.: American Psychological Association.

Celec, Peter, Daniela Ostatnikova, Zdenek Putz, and Matus Kudela. 2002. The Circalunar Cycle of Salivary Testosterone and the Visual-Spatial Performance. *Bratislava Lek Kisty* 103: 59–69.

Chang, K., and R. Antes. 1987. Sex and Cultural Differences in Map Reading. *American Cartographer* 14: 29–42.

Cherrier, M. M., S. Asthana, S. Plymate, L. D. Baker, A. Matsumoto, E. R. Peskind, M. A. Raskind, K. Brodkin, W. Bremner, M. Gonzales, K. Kung, M. Mau, and S. Craft. 2000. Cognitive Effects from Exogenous Manipulation of Testosterone and Estradiol in Older Men. *Society for Neuroscience Abstracts* 30 (11.8). Only available at http://www.sfn.org/index.cfm?pagename=abstracts_ampublications (accessed February, 25, 2008).

Chipman, Karen, and Elizabeth Hampson. 2006. A Female Advantage in the Serial Production of Non-Representational Learned Gestures. *Neuropsychologia* 44: 2315–29.

———, Elizabeth Hampson, and Doreen Kimura. 2002. A Sex Difference in Reliance on Vision During Manual Sequencing Tasks. *Neuropsychologia* 40:910–16.

———, and Doreen Kimura. 1998. An Investigation of Sex Differences on Incidental Memory for Verbal and Pictorial Material. *Learning and Individual Differences* 10: 259–72.

Ciarrochi, Joseph, Keiren Hynes, and Nadia Crittenden. 2005. Can Men Do Better If They Try Harder: Sex and Motivational Effects on Emotional Awareness. *Cognition and Emotion* 19 (1): 133–41.

Clark, A. S., and P. S. Goldman-Rakic. 1989. Gonadal Hormones Influence Emergence of Cortical Function in Nonhuman Primates. *Behavioral Neuroscience* 103 (6): 1287–95.

Cohen-Bendahan, Celina C. C., Cornelieke van de Beek, and Sheri A. Berenbaum. 2005. Prenatal Hormone Effects on Child and Adult Sex-Typed Behavior: Methods and Findings. *Neuroscience and Behavioral Reviews* 29: 353–84.

Colapinto, John. 2004. Gender Gap: What Were the Real Reasons behind David Reimer's Suicide? *Slate*. June 3. http://www.slate.com/id/2101678 (accessed January 20, 2009).

———. 2006. *As Nature Made Him: The Boy Who Was Raised as a Girl (PS)*. New York: Harper Perennial.

Collaer, Marcia L., and Joshua D. Nelson. 2002. Large Visuospatial Sex Differences in Line Judgment: Possible Role of Attentional Factors. *Brain and Cognition* 49: 1–12.

College Board. 2007. Archived SAT data and reports. http://professionals.college-board.com/data-reports-research/sat. pdf (accessed February 25, 2009).

Collins, D. W., and D. Kimura. 1997. A Large Sex Difference on a Two-Dimensional Mental Rotation Task. *Behavioral Neuroscience* 111: 845–49.

Connellan, Jennifer, Simon Baron-Cohen, Sally Wheelwright, Anna Batki, and Jag Ahluwalia. 2001. Sex Differences in Human Neonatal Social Perception. *Infant Behavior and Development* 23: 113–18.

Creighton, H. M. 2004. Long-Term Outcome of Feminization Surgery: The London Experience. *British Journal of Urology International* 93 (supp. 3): 44–46.

Creswell, Catherine S., and David H. Skuse. 1999. Autism in Association with Turner Syndrome: Genetic Implications for Male Vulnerability to Pervasive Developmental Disorders. *Neurocase* 5 (6): 511–18.

Dabbs, James M. 1990. Age and Seasonal Variation in Serum Testosterone Concentration among Men. *Chronobiology International* 7: 245–49.

Dabbs, James M., E-Lee Chang, Rebecca A. Strong, and Rhonda Milun. 1998. Spatial Ability, Navigational Strategy, and Geographical Knowledge among Men and Women. *Evolution and Human Behavior* 19: 89–98.

Daly, Martin, and Margo Wilson. 1983. *Sex, Evolution and Behavior*. 2d ed. Boston: Willard Grant.

Davatzikos, C., and S. M. Resnick. 1998. Sex Differences in Anatomic Measures of Interhemispheric Connectivity: Correlations with Cognition in Women But Not Men. *Cerebral Cortex* 8: 635–40.

Davies, William, Anthony R. Isles, Paul S. Burgoyne, and Lawrence S. Wilkinson. 2006. X-Linked Imprinting: Effects on Brain and Behaviour. *BioEssays* 28 (1): 35–44.

Deary, Ian J., Paul Irwing, Geoff Der, and Timothy C. Bates. 2007. Brother–Sister Differences in the G Factor in Intelligence: Analysis of Full, Opposite-Sex Siblings from NLSY1979. *Intelligence* 35: 451–56.

DeLisi, R., G. Parameswaran, and A. V. McGillicuddy-DeLisi. 1989. Age and Sex Differences in Representation of Horizontality among Children in India. *Perceptual and Motor Skills* 68: 739–46.

Desmond, D. W., D. S. Glenwick, Y. Stern, and T. K. Tatemichi. 1994. Sex Differences in the Representation of Visual Functions in the Human Brain. *Rehabilitation Psychology* 39: 3–14.

Devlin, Anne Sloan. 2004. Sailing Experience and Sex as Correlates of Spatial Ability. *Perceptual and Motor Skills* 98 (3, part 2): 1409–21.

Diamond, Milton, and H. Keith Sigmundson. 1997. Sex Reassignment at Birth: Long Term Review and Clinical Implications. *Archives of Pediatrics and Adolescent Medicine* 151 (10): 298–304.

DiPietro, Janet Ann. 1981. Rough and Tumble Play: A Function of Gender. *Developmental Psychology* 17: 50–58.

Domenice, S., R. V. Corrêa, E. M. F. Costa, M. Y. Nishi, E. Vilain, I. J. P. Arnhold, and B. B. Mendonca. 2003. Mutations in the SRY, DAX1, SF1 and WNT4 Genes in Brazilian Sex-Reversed Patients. *Brazilian Journal of Medical and Biological Research* 36 (12): 145–50.

Donlon, T. F., ed. 1984. *Predictive Validity of the ATP Tests: The College Board Technical Handbook for the Scholastic Aptitude Tests and Achievement Tests.* New York: College Examination Board.

Driesen, Naomi R., and Naftali Raz. 1995. The Influence of Sex, Age, and Handedness on Corpus Callosum Morphology: A Meta-Analysis. *Psychobiology* 23 (3): 240–47.

Ecuyer-Dab, Isabelle, and Michele Robert. 2004. Spatial Ability and Home Range Size: Examining the Relationship in Western Men and Women. *Journal of Comparative Psychology* 118: 217–31.

Ehrhardt, Anke A., G. C. Grisanti, and Heine F. L. Meyer-Bahlburg. 1977. Prenatal Exposure to Medroxyprogesterone Acetate (MPA) in Girls. *Psychoneuroendocrinology* 2: 391–98.

Ekstrom, R. B., J. W. French, H. H. Harmon, and D. Dermen. 1976. *Kit of Factor-Referenced Cognitive Tests.* Princeton, N.J.: Educational Testing Service.

Engelhard, George. 1990. Gender Differences in Performance on Mathematics Items: Evidence from the United States and Thailand. *Contemporary Educational Psychology* 15: 13–26.

Evans, E. Margaret, Heidi Schweingruber, and Harold W. Stevenson. 2002. Gender Differences in Interest and Knowledge Acquisition: The United States, Taiwan, and Japan. *Sex Roles* 47 (3/4): 153–67.

Feingold, Alan. 1988. Cognitive Gender Differences are Disappearing. *American Psychologist* 43: 95–103.

———. 1992. Sex Differences in Variability in Intellectual Abilities: A New Look at an Old Controversy. *Review of Educational Research* 62 (1): 61–84.

———. 1996. Cognitive Gender Differences: Where Are They and Why Are They There? *Learning and Individual Differences* 8: 25–32.

Feldman, S. Shirley, Sharon Nash, and C. Cutrona. 1977. The Influence of Age and Sex on Responsiveness to Babies. *Developmental Psychology* 13: 675–76.

Felson, Richard B., and Lisa Trudeau. 1991. Gender Differences in Mathematical Performance. *Social Psychology Quarterly* 54: 113–26.

Ferin, Michel, Raphael Jewelewicz, and Michelle Warren. 1993. *The Menstrual Cycle: Physiology, Reproductive Disorders, and Infertility.* New York: Oxford University Press.

Ford, Martin. 1982. Social Cognition and Social Competence in Adolescence. *Developmental Psychology* 18: 323–40.

Frome, Pamela M., Corinne J. Alfield, Jacquelynne S. Eccles, and Bonnie L. Barber. 2006. Why Don't They Want a Male-Dominated Job? An Investigation of Young Women Who Changed Their Occupational Aspirations. *Educational Research and Evaluation* 12 (4): 359–72.

Galea, Liisa A. M., and Doreen Kimura. 1993. Sex Differences in Route Learning. *Personality and Individual Differences* 14: 53–65.

Gamé, Florence, Isabelle Carchon, and François Vital-Durand. 2003. The Effect of Stimulus Attractiveness on Visual Tracking in 2- to 6-Month-Old Infants. *Infant Behavior and Development* 26: 135–50.

Gannon, Frank, Sara Quirk, and Sebastian Guest. 2001. Searching for Discrimination: Are Women Treated Fairly in the EMBO Postdoctoral Fellowship Scheme? *EMBO Reports* 2 (8): 655–57.

Gartler, S. M., P. Andina, and N. Gant. 1975. Ontogeny of X-Chromosome Inactivation in the Female Germ Line. *Experimental Cell Research* 91: 454–57.

Gartler, S. M., R. M. Liskay, B. K. Campbell, R. Sparkes, and N. Gant. 1972. Evidence for Two Functional X Chromosomes in Human Oocytes. *Cell Differentiation* 1: 215–18.

Geary, David C., and M. Catherine DeSoto. 2001. Sex Differences in Spatial Abilities among Adults from the United States and China. *Evolution and Cognition* 7: 172–77.

Good, Catriona D., Kate Lawrence, N. Simon Thomas, Cathy J. Price, John Ashburner, Karl J. Friston, Richard S. J. Frackowiak, Lars Oreland, and David H. Skuse. 2003. Dosage-Sensitive X-Linked Locus Influences the Development of the Amygdala and Orbitofrontal Cortex and Fear Recognition in Humans. *Brain* 126: 2431–46.

Gouchie, Catherine, and Doreen Kimura. 1991. The Relationship between Testosterone Levels and Cognitive Ability Patterns. *Psychoneuroendocrinology* 16:323–34.

Gouzoules, Henry, and Sarah Gouzoules. 1989. Sex Differences in the Acquisition of Communicative Competence by Pigtail Macaques (Macaca-Nemestrina). *American Journal of Primatology* 19: 163–74.

Goy, R. W., F. B. Bercovitch, and M. C. McBrair. 1988. Behavioral Masculinization is Independent of Genital Masculinization in Prenatally Androgenized Female Rhesus Macaques. *Hormones and Behavior* 22: 552–71.

Goy, R. W., and Bruce S. McEwen. 1980. *Sexual Differentiation of the Brain*. Cambridge, Mass.: MIT Press.

Goy, R. W., and J. A. Resko. 1972. Gonadal Hormones and Behavior of Normal and Pseudohermaphroditic Nonhuman Female Primates. *Recent Progress in Hormone Research* 28: 707–33.

Gunnar, Megan R., and Margaret Donahue. 1980. Sex Differences in Social Responsiveness between Six Months and Twelve Months. *Child Development* 51 (1): 262–65.

Hagger, C., J. Bachevalier, and B. B. Bercu. 1987. Sexual Dimorphism in the Development of Habit Formation: Effects of Perinatal Steroid Hormones. *Neuroscience* 22 (supp.): S520.

Hall, Jeffry A., and Doreen Kimura. 1995. Sexual Orientation and Performance on Sexually Dimorphic Motor Tasks. *Archives of Sexual Behavior* 24: 395–407.

Hall, Judith A. 1984. *Nonverbal Sex Differences: Communication Accuracy and Expressive Style*. Baltimore, Md.: Johns Hopkins University Press.

Halpern, Diane F., Camilla P. Benbow, David C. Geary, Ruben C. Gur, Janet Shibley Hyde, and Morton Ann Gernsbacher. 2007. The Science of Sex Differences in Science and Mathematics. *Psychological Science in the Public Interest* 8:1–51.

Hampson, Elizabeth. 1990. Variations in Sex-Related Cognitive Abilities Across the Menstrual Cycle. *Brain and Cognition* 14: 26–43.

Hampson, Elizabeth, Joanne F. Rovet, and Deborah Altmann. 1998. Spatial Reasoning in Children with Congenital Adrenal Hyperplasia Due to 21-Hydroxylase Deficiency. *Developmental Neuropsychology* 14: 299–320.

Hampson, Elizabeth, Sari M. van Anders, and Lucy I. Mullin. 2006. A Female Advantage in the Recognition of Emotional Facial Expressions: Test of an Evolutionary Hypothesis. *Evolution and Human Behavior* 27 (6): 401–16.

Hampson, Elizabeth, and Elizabeth A. Young. 2008. Methodological Issues in the Study of Hormone-Behavior Relations in Humans: Understanding and Monitoring the Menstrual Cycle. In *Sex Differences in the Brain from Genes to Behavior*, ed. Jill B. Becker, Karen J. Berkley, Nori Geary, Elizabeth Hampson, James P. Herman, and Elizabeth A. Young, 63–78. New York: Oxford University Press.

Harshman, R. A., E. Hampson, and S. A. Berenbaum. 1983. Individual Differences in Cognitive Abilities and Brain Organization, Part I: Sex and Handedness Differences in Ability. *Canadian Journal of Psychology* 37: 144–92.

Hassett, Janice M., Erin R. Siebart, and Kim Wallen. 2004. Sexually Differentiated Toy Preferences in Rhesus Monkeys [Abstract]. *Hormones and Behavior* 46: 91.

Hausmann, Markus, Ditte Slabbekoorn, Stephanie H. M. Van Goozen, Peggy T. Cohen-Kettenis, and Onur Gunterkun. 2000. Sex Hormones Affect Spatial Abilities during the Menstrual Cycle. *Behavioral Neuroscience* 114: 1245–50.

Hedges, Larry V., and Amy Nowell. 1995 Sex Differences in Mental Test Scores, Variability, and Numbers of High-Scoring Individuals. *Science* 269: 41–45.

Helleday, Jan, Aniko Bartfai, E. Martin Ritzén, and Marianne Forsman. 1994. General Intelligence and Cognitive Profile in Women with Congenital Adrenal Hyperplasia (CAH). *Psychoneuroendocrinology* 19 (4): 343–56.

Herlitz, A., E. Airaksinen, and E. Nordström. 1999. Sex Differences in Episodic Memory: The Impact of Verbal and Visuospatial Memory. *Neuropsychology* 13: 590–97.

Herman, Rebecca A., Megan A. Measday, and Kim Wallen. 2003. Sex Differences in Interest in Infants in Juvenile Rhesus Monkeys: Relationship to Prenatal Androgen. *Hormones and Behavior* 43: 573–83.

Hier, D. B., W. B. Yoon, J. P. Mohr, T. R. Price, and P. A. Wolf. 1994. Gender and Aphasia in the Stroke Data Bank. *Brain and Language* 47: 155–67.

Hines, Melissa. 2004. *Brain Gender.* New York: Oxford University Press.

Hines, Melissa, S. Faisal Ahmed, and Ieuan A. Hughes. 2003. Psychological Outcomes and Gender-Related Development in Complete Androgen Insensitivity Syndrome. *Archives of Sexual Behavior* 32 (2): 93–101.

Hines, Melissa, Charles Brook, and Gerard S. Conway. 2004. Androgen and Psychosexual Development: Core Gender Identity, Sexual Orientation and Recalled Childhood Gender Role Behavior in Men and Women with Congenital Adrenal Hyperplasia (CAH). *Journal of Sex Research* 41: 75–81.

Hines, Melissa, Briony A. Fane, Vickie L. Pasterski, Greta A. Mathews, Gerard S. Conway, and Charles Brook. 2003. Spatial Abilities Following Prenatal Androgen Abnormality: Targeting and Mental Rotations Performance in Individuals with Congenital Adrenal Hyperplasia. *Psychoneuroendocrinology* 28: 1010–26.

Hines, Melissa, Susan Golombok, John Rust, Katie Johnston, and Jean Golding. 2002. The ALSPAC Study Team. Testosterone During Pregnancy and Childhood Gender Role Behavior: A Longitudinal Population Study. *Child Development* 73: 1678–87.

Hittelman, Joan, and Robert Dickes. 1979. Sex Differences in Neonatal Eye Contact Time. *Merrill-Palmer Quarterly* 25 (3): 171–84.

Holding, C. S., and D. H. Holding. 1989. Acquisition of Route Network Knowledge by Males and Females. *Journal of General Psychology* 116: 29–41.

Hugdahl, Kenneth, Tormod Thomsen, and Lars Ersland. 2006. Sex Differences in Visuo-Spatial Processing: An fMRI Study of Mental Rotation. *Neuropsychologia* 44: 1575–83.

Iijima, Megumi, Osamu Arisaka, Fumie Minamoto, and Yasumasa Arai. 2001. Sex Differences in Children's Free Drawings: A Study on Girls with Congenital Adrenal Hyperplasia. *Hormones and Behavior* 40: 99–104.

Imperato-McGinley, Jullianne, Marino Pichardo, Teofilo Gautier, Daniel Voyer, and M. Philip Bryden. 1991. Cognitive Abilities in Androgen-Insensitive Subjects: Comparison with Control Males and Females of the Same Kindred. *Clinical Endocrinology* 34: 341–47.

Imperato-McGinley, Jullianne, and Yuan-Shan Zhu. 2002. Androgens and Male Physiology: The Syndrome of 5'-Reductase-2 Deficiency. *Molecular and Cellular Endocrinology* 198 (1–2): 51–59.

Ingram, D. 1975. Motor Asymmetries in Young Children. *Neuropsychologia* 13: 95–102.

International Mathematical Olympiad. 2008. Hall of Fame. http://www.imo-official.org/hall.aspx (accessed March 4, 2009).

Isgor, Ceylan, and Dale R. Sengelaub. 1998. Prenatal Gonadal Steroids Affect Adult Spatial Behavior, CA1 and CA3 Pyramidal Cell Morphology in Rats. *Hormones and Behavior* 34: 183–98.

———. 2003. Effects of Neonatal Gonadal Steroids on Adult CA3 Pyramidal Neuron Dendritic Morphology and Spatial Memory in Rats. *Journal of Neurobiology* 55: 179–90.

Jackson, Douglas N., and J. Philippe Rushton. 2006. Males Have Greater G: Sex Differences in General Mental Ability from 100,000 17- to 18-Year Olds on the Scholastic Assessment Test. *Intelligence* 34: 479–86.

Janowsky, J. S., S. K. Oviatt, and E. S. Orwoll. 1994. Testosterone Influences Spatial Cognition in Older Men. *Behavioral Neuroscience* 108: 325–32.

Jardine, Rosemary, and N. G. Martin. 1983. Spatial Ability and Throwing Accuracy. *Behavior Genetics* 13: 331–40.

Jensen, Arthur Robert. 1988. Sex Differences in Arithmetic Computation and Reasoning in Prepubertal Boys and Girls. *Behavioral and Brain Sciences* 11: 198–99.

Jobson, Steve, and John S. Watson. 1984. Sex and Age Differences in Choice Behaviour: The Object–Person Dimension. *Perception* 13 (6): 719–24.

Johnsson-Smaragdi, Ulla, and Annelis Jönsson. 2006. Book Reading in Leisure Time: Long-Term Changes in Young People's Book Reading Habits. *Scandinavian Journal of Educational Research* 50 (5): 519–40.

Jordan, Kristen, Torsten Wustenberg, Hans-Jochen Heinze, Michael Peters, and Lutz Jancke. 2002. Women and Men Exhibit Different Cortical Activation Patterns During Mental Rotation Tasks. *Neuropsychologia* 40: 2397–2408.

Jürgensen, Martina, Olaf Hiort, Paul-Martin Holterhus, and Ute Thyen. 2007. Gender Role Behavior in Children with XY Karyotype and Disorders of Sex Development. *Hormones and Behavior* 51: 443–53.

Kail, Robert V., and Alexander W. Siegel. 1977. Sex Differences in Retention of Verbal and Spatial Characteristics of Stimuli. *Journal of Experimental Child Psychology* 23: 341–47.

Kimura, Doreen. 1983. Sex Differences in Cerebral Organization for Speech and Praxic Functions. *Canadian Journal of Psychology* 37: 19–35.

———. 1987. Are Men's and Women's Brains Really Different? *Canadian Psychology* 28: 133–47.

———. 1994. Body Asymmetry and Intellectual Pattern. *Personality and Individual Differences* 17: 53–60.

———. 1997. *Neuropsychology Test Procedures.* Revised ed. London, Canada: D. K. Consultants.

———. 1999. *Sex and Cognition.* Cambridge, Mass.: MIT Press.

———. 2006. Recollections of an Accidental Contrarian. *Canadian Journal of Psychology* 60: 80–89.

Kimura, Doreen, and Paul G. Clarke. 2002. Women's Advantage on Verbal Memory Is Not Restricted to Concrete Words. *Psychological Reports* 91: 1137–42.

Kimura, Doreen, and Elizabeth Hampson. 1994. Cognitive Pattern in Men and Women Is Influenced by Fluctuations in Sex Hormones. *Current Directions in Psychological Science* 3: 57–61.

Kimura, Doreen, and R. A. Harshman. 1984. Sex Differences in Brain Organization for Verbal and Non-Verbal Functions. In *Sex Differences in the Brain: Progress in Brain*

Research, ed. G. J. DeVries, J. P. C. DeBruin, H. B. M. Uylings, and M. A. Corner, 423–39. Amsterdam: Elsevier.

Kobrin, Jennifer L., and Ernest W. Kimmel. 2006. Test Development and Technical Information on the Writing Section of the SAT Reasoning Test. http://professionals.collegeboard.com/data-reports-research/cb/test-development-sat-writing.pdf (accessed March 4, 2009).

Komnenich, Pauline, David M. Lane, Richard P. Dickey, and Sergio C. Stone. 1978. Gonadal Hormones and Cognitive Performance. *Physiological Psychology* 6: 115–20.

LaFreniere, Peter J., and Jean E. Dumas. 1996. Social Competence and Behavior Evaluation in Children Ages 3 to 6 Years: The Short Form (SCBE–30). *Psychological Assessment* 8 (4): 369–77.

Lasco, Mitchell S., Theresa J. Jordana, Mark A. Edgar, Carol K. Petito, and William Bynee. 2002. A Lack of Dimorphism of Sex or Sexual Orientation in the Human Anterior Commissure. *Brain Research* 936: 95–98.

Law, David J., James W. Pellegrino, and Earl B. Hunt. 1993. Comparing the Tortoise and the Hare: Gender Differences and Experience in Dynamic Spatial Reasoning Tasks. *Psychological Science* 4: 35–40.

Lee, Richard E., and Irven DeVore, eds. 1968. *Man the Hunter*. Chicago: Aldine.

Levine, R. J. 1991. Seasonal Variation in Human Semen Quality. In *Temperature and Environmental Effects on the Testis*, ed. A. W. Zorgnatti, 89–96. New York: Plenum.

Levy, J., and W. Heller. 1992. Gender Differences in Human Neuropsychological Function. In *Handbook of Behavioral Neurobiology*, Volume 11, Sexual Differentiation, ed. A. A. Gerall, H. Moltz, and I. L. Ward, 245–74. New York: Plenum Press.

Levy, J., and C. Trevarthen. 1977. Perceptual, Semantic and Phonetic Aspects of Elementary Language Processes in Split-Brain Patients. *Brain* 100: 105–18

Lewin, Catharina, G. Wolgers, and A. Herlitz. 2001. Sex Differences Favoring Women in Verbal But Not in Visuospatial Episodic Memory. *Neuropsychology* 15: 165–73.

Lewis, Charles, and Warren W. Willingham. 1995. *The Effects of Sample Restriction on Gender Differences*. Research Report ETS RR-95-13. Princeton, N.J.: Educational Testing Service.

Ligon, Ernest Mayfield. 1932. A Genetic Study of Color Naming and Word Reading. *American Journal of Psychology* 44: 103–22.

Lindeman, Marjaana. 1991. Sex Differences in Discriminating Social Information. *Journal of Social Behavior and Personality* 6 (2): 229–36.

Lovejoy, C. Owen. 1981. The Origin of Man. *Science* 211: 341–50.

Low, Renae, and Ray Over. 1993. Gender Differences in Solutions of Algebraic Word Problems Containing Irrelevant Information. *Journal of Educational Psychology* 85: 331–39.

Lubinski, David, and Camilla P. Benbow. 1992. Gender Differences in Abilities and Preferences among the Gifted: Implications for the Math-Science Pipeline. *Current Directions in Psychological Science* 1: 61–66.

————. 2006. Study of Mathematically Precocious Youth after 35 Years. *Perspectives on Psychological Science* 1 (4): 316–45.

Lubinski, David, Camilla P. Benbow, Rose Mary Webb, and April Bleske-Rechek. 2006. Tracking Exceptional Human Capital over Two Decades. *Psychological Science* 17 (3): 194–99.

Lummis, Max, and Harold W. Stevenson. 1990. Gender Differences in Beliefs and Achievement: A Cross-Cultural Study. *Developmental Psychology* 26: 254–63.

Lutchmaya, Svetlana, and Simon Baron-Cohen. 2002. Human Sex Differences in Social and Non-Social Looking Preferences, at 12 Months of Age. *Infant Behavior and Development* 25: 319–25.

Lutchmaya, Svetlana, Simon Baron-Cohen, and Peter Raggatt. 2002. Foetal Testosterone and Eye Contact in 12-Month-Old Human Infants. *Infant Behavior and Development* 25: 327–35.

Lynn, Richard. 1994. Sex Differences in Intelligence and Brain Size: A Paradox Resolved. *Personality and Individual Differences* 17: 257–71.

Lynn, Richard, and Paul Irwing. 2002. Sex Differences in General Knowledge, Semantic Memory, and Reasoning Ability. *British Journal of Psychology* 93: 545–56.

MacArthur, Russell. 1967. Sex Differences in Field Independence for the Eskimo. *International Journal of Psychology* 2: 139–40.

Machin, Stephen, and Tuomas Pekkarinen. 2008. Global Sex Differences in Test Score Variability. *Science* 322: 1331–32. Supporting online material at www.sciencemag.org/cgi/content/full/322/5906/1331/DC1.pdf (accessed March 5, 2009).

MacLusky, Neil J., and Frederick Naftolian. 1981. Sexual Differentiation of the Central Nervous System. *Science* 211: 1294–1303.

Maestripieri, Dario, and Suzanne Pelka. 2002. Sex Differences in Interest in Infants across the Lifespan: A Biological Adaptation for Parenting? *Human Nature* 13: 327–44.

Malouf, Matthew A., Claude J. Migeon, Kathryn A. Carson, Loredana Petrucci, and Amy B. Wisniewski. 2006. Cognitive Outcome in Adult Women Affected by Congenital Adrenal Hyperplasia Due to 21-Hydroxylase Deficiency. *Hormone Research* 65: 142–50.

Mann, V. A., S. Sasanuma, N. Sakuma, and S. Masaki. 1990. Sex Differences in Cognitive Abilities: A Cross-Cultural Perspective. *Neuropsychologia* 28:1063–77.

Marco, Elysa J., and David H. Skuse. 2006. Autism-Lessons from the X Chromosome. *Social Cognitive and Affective Neuroscience* 1 (3): 183–93.

Marshall, S. P., and J. D. Smith. 1987. Sex Differences in Learning Mathematics: A Longitudinal Study with Item and Error Analyses. *Journal of Educational Psychology* 79: 372–83.

Masica, D. N., J. Money, A. A. Ehrhardt, and V. G. Lewis. 1969. I.Q., Fetal Sex Hormones and Cognitive Patterns: Studies in the Testicular Feminizing Syndrome of Androgen Insensitivity. *Johns Hopkins Medical Journal* 124: 34–43.

Masters, Mary Soares, and Barbara Sanders. 1993. Is the Gender Difference in Mental Rotation Disappearing? *Behavior Genetics* 23: 337–41.

Mateer, C., S. B. Polen, and G. Ojemann. 1982. Sexual Variation in Cortical Localization of Naming as Determined by Stimulation Mapping. *Behavioral and Brain Sciences* 5: 310–11.

Mathematical Association of America. 2008. The Mathematical Association of America's William Lowell Putnam Competition. http://www.maa.org/awards/putnam.html (accessed February 25, 2009).

Matzuk, Martin M., Kathleen H. Burns, Maria M. Viveiros, and John J. Eppig. 2002. Intercellular Communication in the Mammalian Ovary: Oocytes Carry the Conversation. *Science* 296: 2178–80.

Mayer, A., G. Lahr, D. F. Swaab, C. Pilgrim, and I. Reisert. 1998. The Y-Chromosome Genes SRY and ZFY Are Transcribed in Adult Human Brain. *Neurogenetics* 1: 281–88.

Mayes, J. T., G. Jahoda, and I. Neilson. 1988. Patterns of Visual-Spatial Performance and "Spatial Ability": Dissociation of Ethnic and Sex Differences. *British Journal of Psychology* 79: 105–19.

McBurney, D. H., S. J. C. Gaulin, T. Devineni, and C. Adams. 1997. Superior Spatial Memory of Women: Stronger Evidence for the Gathering Hypothesis. *Evolution and Human Behavior* 18: 165–74.

McGlone, J., and A. J. Fox. 1982. Evidence from Sodium Amytal Studies of Greater Asymmetry of Verbal Representation in Men Compared to Women. In *Advances in Epileptology. XIIIth Epilepsy International Symposium*, ed. H. Akimoto, H. Kazamatsuri, M Seino, and A. Ward, 389–91. New York: Raven Press.

McGuinness, D., and J. Sparks. 1983. Cognitive Style and Cognitive Maps: Sex Differences in Representations of a Familiar Terrain. *Journal of Mental Imagery* 7: 91–100.

McLaren, Anne. 1991. Development of the Mammalian Gonad: The Fate of the Supporting Cell Lineage. *BioEssays* 13: 151–56.

Meikle, A. W., D. T. Bishop, J. D. Stringham, and D. W. West. 1987. Quantitating Genetic and Nongenetic Factors that Determine Plasma Sex Steroid Variation in Normal Male Twins. *Metabolism* 35: 1090–95.

Melo, Karla F. S., Berenice B. Mendonca, Ana Elisa C. Billerbeck, Elaine M. F. Costa, Marlene Inacio, Frederico A. Q. Silva, Angela M. O. Leal, Ana C. Latronico, Ivo J. P. Arnhold. 2003. Clinical, Hormonal, Behavioral, and Genetic Characteristics of Androgen Insensitivity Syndrome in a Brazilian Cohort: Five Novel Mutations in the Androgen Receptor Gene. *Journal of Clinical Endocrinology and Metabolism* 88 (7): 3241–50.

Miller, L. K., and V. Santoni. 1986. Sex Differences in Spatial Abilities: Strategic and Experiential Correlates. *Acta Psychologica* 62: 225–35.

Mitchell, Gregory, and Philip E. Tetlock. 2006. Antidiscrimination Law and the Perils of Mindreading. *Ohio State Law Journal* 67 (5): 1023–1121.

Moffat, Scott D., and Elizabeth Hampson. 1996. A Curvilinear Relationship between Testosterone and Spatial Cognition in Humans: Possible Influence of Hand Preference. *Psychoneuroendocrinology* 21: 323–37.

Moffat, Scott D., Elizabeth Hampson, and M. Hatzipantelis. 1998. Navigation in a "Virtual" Maze: Sex Differences and Correlation with Psychometric Measures of Spatial Ability in Humans. *Evolution and Human Behavior* 19: 73–87.

Moore, Elise G. J., and A. Wade Smith. 1987. Sex and Ethnic Group Differences in Mathematics Achievement: Results from the National Longitudinal Study. *Journal for Research in Mathematics Education* 18: 25–36.

Mueller, S. C., V. Temple, E. Ohb, C. VanRyzin, A. Williams, B. Cornwell, C. Grillon, D. S. Pine, M. Ernst, and D. P. Merke. 2008. Early Androgen Exposure Modulates Spatial Cognition in Congenital Adrenal Hyperplasia (CAH). *Psychoneuroendocrinology* 33: 973–80.

Muller, Carol B., Sally M. Ride, Janie Fouke, Telle Whitney, Denice D. Denton, Nancy Cantor, Donna J. Nelson, Jim Plummer, et al. 2005. Gender Differences and Performance in Science. *Science* 307 (5712): 1043.

Munroe, R. L., and R. H. Munroe. 1971. Effect of Environmental Experience on Spatial Ability in an East African Society. *Journal of Social Psychology* 83: 15–22.

Murphy, Laura O., and Steven M. Ross. 1987. Gender Differences in the Social Problem-Solving Performance of Adolescents. *Sex Roles* 16 (5/6): 251–64.

National Academy of Sciences, National Academy of Engineering, and Institute of Medicine of the National Academies. 2007. *Beyond Bias and Barriers: Fulfilling the Potential of Women in Academic Science and Engineering.* Washington, D.C.: National Academies Press.

Newcombe, N. S. 2006. Taking Science Seriously: Straight Thinking about Spatial Sex Differences. In *Why Aren't More Women in Science?* ed. Stephen J. Ceci and Wendy M. Williams, 69–77. Washington, D.C.: American Psychological Association.

Nicholson, Karen G., and Doreen Kimura. 1996. Sex Differences for Speech and Manual Skill. *Perceptual and Motor Skills* 82: 3–13.

Nordenstrom, Anna, Anna Servin, Gunilla Bohlin, Agne Larsson, and Anna Wedel. 2002. Sex-Typed Toy Play Behavior Correlates with the Degree of Prenatal Androgen Exposure Assessed by CYP21 Genotype in Girls with Congenital Adrenal Hyperplasia. *Journal of Clinical Endocrinology and Metabolism* 87 (11): 5119–24.

Nowell, Amy, and Larry V. Hedges. 1998. Trends in Gender Differences in Academic Achievement from 1960 to 1994: An Analysis of Differences in Mean, Variance, and Extreme Scores. *Sex Roles* 39 (1/2): 21–43.

Olafsen, Kåre S., John A. Rønning, Per Ivar Kaaresen, Stein Erik Ulvund, Bjørn Helge Handegård, and Lauritz Bredrup Dahl. 2006. Joint Attention in Term and Preterm Infants at 12 Months Corrected Age: The Significance of Gender and Intervention Based on a Randomized Controlled Trial. *Infant Behavior and Development* 29: 554–63.

Olszewski-Kubilius, Paula, and Dana Turner. 2002. Gender Differences among Elementary School-Aged Gifted Students in Achievement, Perceptions of Ability, and Subject Preference. *Journal for the Education of the Gifted* 25 (3): 233–68.

Ostatníková, Daniela, Zdenek Putz, and P. Matejka. 1995. Circannual Fluctuations of Free Testosterone Concentrations in Men and Women. Abstract. *Physiological Research* 44 (4): 19. Available only at http://www.testosterone.host.sk/Publik.htm#1995 (accessed March 5, 2009).

Ostatníková, Daniela, Zdenek Putz, P. Matûjka, and M. Országh. 1995. Kolísanie hladiny testosterónu v slinách mužov a žien v priebehu [Circannual fluctuations of salivary testosterone concentrations in men and women]. *Praktická Gynekológia* 2: 45–48.

Ostatníková, Daniela, Zdenek Putz, Peter Celec, and Július Hodosy. 2002. May Testosterone Levels and Their Fluctuations Influence Cognitive Performance in Humans? *Scripta Medica (Brno)* 75 (5): 245–54.

Ottolenghi, Chris, Maria Colombino, Laura Crisponi, Antonio Cao, Antonino Forabosco, David Schlessinger, and Manuela Uda. 2007. Transcriptional Control of Ovarian Development in Somatic Cells. *Seminars in Reproductive Medicine* 25 (4): 252–63.

Owen, Ken, and Richard Lynn. 1993. Sex Differences in Primary Cognitive Abilities among Blacks, Indians and Whites in South Africa. *Journal of Biosocial Science* 25: 557–60.

Park, Gregory, David Lubinski, and Camilla P. Benbow. 2007. Contrasting Intellectual Patterns Predict Creativity in the Arts and Sciences. *Psychological Science* 18 (11): 948–52.

Pasterski, Vickie L., Peter Hindmarsh, Mitchell E. Geffner, Charles Brook, Caroline Brain, and Melissa Hines. 2007. Increased Aggression and Activity Level in 3- to 11-Year-Old Girls with Congenital Adrenal Hyperplasia (CAH). *Hormones and Behavior* 52: 368–74

Peters, Michael, Wolfgang Lehmann, Sayuri Takahira, Yoshiaki Takeuchi, and Kirsten Jordan. 2006. Mental Rotation Test Performance in Four Cross-Cultural Samples (N=3367): Overall Sex Differences and the Role of Academic Program in Performance. *Cortex* 42: 1005–14.

Peterson, R. C. 1993. A Sex Difference in Stereoscopic Depth Perception. Master's thesis, Department of Psychology, University of Western Ontario, London, Ontario, Canada.

Podrouzek, Wayne, and David Furrow. 1988. Preschoolers' Use of Eye Contact while Speaking: The Influence of Sex, Age, and Conversational Partner. *Journal of Psycholinguistic Research* 17 (2): 89–98.

Pontius, Anneliese A. 1997a. Lack of Sex Differences among East Ecuadorian School Children on Geometric Figure Rotation and Figure Drawing. *Perceptual and Motor Skills* 85: 72–74.

———. 1997b. No Gender Difference in Spatial Representation by School Children in Northwest Pakistan. *Journal of Cross-Cultural Psychology* 28: 779–86.

Postma, Albert, Joke Winkel, Adriaan Tuiten, and Jack van Honk. 1999. Sex Differences and Menstrual Cycle Effects in Human Spatial Memory. *Psychoneuroendocrinology* 24: 175–92.

Proverbio, Alice Mado, Silvia Matarazzo, Valentina E. Brignone, Marzia Del Zotto, and Alberto Zani. 2007. Processing Valence and Intensity of Infant Expressions: The Roles of Expertise and Gender. *Scandinavian Journal of Psychology* 48: 477–85

Puts, David A., Michael A. McDaniel, Cynthia L. Jordan, and Marc S. Breedlove. 2008. Spatial Ability and Prenatal Androgens: Meta-Analyses of Congenital Adrenal Hyperplasia and Digit Ratio (2d:4d) Studies. *Archives of Sexual Behavior* 37 (1): 100–111.

Reinberg, Alain, and Michel Lagoguey. 1978. Circadian and Circannual Rhythms in Sexual Activity and Plasma Hormones (FSH, LH, Testosterone) of Five Human Males. *Archives of Sexual Behavior* 7: 13–30.

Reiner, William G., and John P. Gearhart. 2004. Discordant Sexual Identity in Some Genetic Males with Cloacal Exstrophy Assigned to Female Sex at Birth. *New England Journal of Medicine* 350 (4): 333–41.

Resnick, S. M., S. A. Berenbaum, I. I. Gottesmann, and T. J. Bouchard. 1986. Early Hormonal Influences on Cognitive Functioning in Congenital Adrenal Hyperplasia. *Developmental Psychology* 22: 191–98.

Reynaud, Karine, Rita Cortvrindt, Franciska Verlinde, Jean De Schepper, Claire Bourgain, and Johan Smitz. 2004. Number of Ovarian Follicles in Human Fetuses With the 45,X Karyotype. *Fertility and Sterility* 81 (4): 1112–19.

Robert, M., and F. Harel. 1996. The Gender Difference in Orienting Liquid Surfaces and Plumb Lines: Its Robustness, Its Correlates, and the Associated Knowledge of Simple Physics. *Canadian Journal of Experimental Psychology* 50: 280–314.

Robert, M., and Théophile Ohlmann. 1994. Water-Level Representation by Men and Women as a Function of Rod-and-Frame Test Proficiency and Visual and Postural Information. *Perception* 23 (11): 1321–33

Rogers, H. H., and B. C. J. Hamel. 2005. X-Linked Mental Retardation. *Nature Reviews Genetics* 6 (January): 46–57.

Roof, Robin L. 1993. Neonatal Exogenous Testosterone Modifies Sex Difference in Radial Arm and Morris Water Maze Performance in Prepubescent and Adult Rats. *Behavioral Brain Research* 53: 1–10.

Rosenthal, Robert, Judith A. Hall, M. Robin DiMatteo, Peter L. Rogers, and Dane Archer. 1979. *Sensitivity to Nonverbal Communication: The PONS Test.* Baltimore: Johns Hopkins University Press.

Rushton, J. P., and C. D. Ankney. 2007. The Evolution of Brain Size and Intelligence. In *Evolutionary Cognitive Neuroscience*, ed. S. M. Platek, J. P. Keenan, and T. K. Shackleford, 121–61. Cambridge, Mass.: MIT Press.

Salat, D., A. Ward, J. A. Kaye, and J. S. Janowsky. 1997. Sex Differences in the Corpus Callosum with Aging. *Neurobiology of Aging* 18 (2): 191–97.

Sanders, G., K. Sinclair, and T. Walsh. 2007. Testing Predictions from the Hunter-Gatherer Hypothesis—2: Sex Differences in the Visual Processing of Near and Far Space. *Evolutionary Psychology* 5: 666–79.

Sandström, C., and I. Lundberg. 1956. A Genetic Approach to Sex Differences in Localization. *Acta Psychologica* 12: 247–53.

Saucier, Deborah M., Sheryl M. Green, Jennifer Leason, Alastair MacFadden, Scott Bell, and Lorin J. Elias. 2002. Are Sex Differences in Navigation Caused by Sexually Dimorphic Strategies or by Differences in the Ability to Use the Strategies? *Behavioral Neuroscience* 116: 403–10.

Saucier, Deborah M., and Doreen Kimura. 1998. Intrapersonal Motor But Not Extrapersonal Targeting Skill is Enhanced during the Midluteal Phase of the Menstrual Cycle. *Developmental Neuropsychology* 14: 385–98.

Saucier, Deborah M., Amanda Lisoway, Sheryl Green, and Lorin Elias. 2007. Female Advantage for Object Location Memory in Peripersonal But Not Extrapersonal Space. *Journal of the International Neuropsychological Society* 13: 683–86.

Schirmer, Annett, Tricia Striano, and Angela Friederici. 2005. Sex Differences in the Preattentive Processing of Vocal Emotional Expressions. *NeuroReport* 16 (6): 635–39.

Shaywitz, Bennett, Sally E. Shaywitz, Kenneth R. Pugh, R. Todd Constable, Pawel Skudlarski, Robert K. Fulbright, Richard A. Bronen, Jack M. Fletcher, Donald P. Shankweiler, Leonard Katz, and John C. Gores. 1995. Sex Differences in the Functional Organization of the Brain for Language. *Nature* 373: 607–9.

Shea, Daniel L., David Lubinski, and Camilla P. Benbow. 2001. Importance of Assessing Spatial Ability in Intellectually Talented Young Adolescents: A 20-year Longitudinal Study. *Journal of Educational Psychology* 93: 604–14.

Shute, Valerie J., James W. Pellegrino, Lawrence Hubert, and Robert W. Reynolds. 1983. The Relationship between Androgen Levels and Human Spatial Abilities. *Bulletin of the Psychonomic Society* 21: 465–68.

Silverman, I., and M. Eals. 1992. Sex Differences in Spatial Abilities: Evolutionary Theory and Data. In *The Adapted Mind*, ed. J. H. Barkow, L. Cosmides, and J. Tooby, 533–49. New York: Oxford.

Silverman, I., and Krista Phillips. 1993. Effects of Estrogen Changes during the Menstrual Cycle on Spatial Performance. *Ethology and Sociobiology* 14: 257–70.

Simpson, Joe Leigh, and Aleksandar Rajkovic. 1999. Ovarian Differentiation and Gonadal Failure. *American Journal of Medical Genetics (Seminars in Medical Genetics)* 89: 186–200.

Skuse, D. H., R. S. James, D. V. M. Bishop, B. Coppins, P. Dalton, G. Aamodt-Leeper, M. Bacarese-Hamilton, C. Creswell, R. McGurk, and P. A. Jacobs. 1997. Evidence from Turner's Syndrome of an Imprinted X-linked Locus Affecting Cognitive Function. *Nature* 387: 705–8.

Slabbekoorn, Ditte, Stephanie H. M. Van Goozen, Jos Megens, Louis J. G. Gooren, and Peggy T. Cohen-Kettenis. 1999. Activating Effects of Cross-Sex Hormones on Cognitive Functioning: A Study of Short-Term and Long-Term Hormone Effects in Transsexuals. *Psychoneuroendocrinology* 24: 423–47.

Smals, A. G. H., P. W. C. Kloppenborg, and T. H. Benrad. 1976. Circannual Cycles in Plasma Testosterone Levels in Man. *Journal of Clinical Endocrinology and Metabolism* 42: 979–82.

Smith, A. 1967. Consistent Sex Differences in a Specific (Decoding) Test Performance. *Educational and Psychological Measurement* 27: 1077–83.

Spellacy, W. N., I. C. Bernstein, and W. H. Cohen. 1965. Complete Form of Testicular Feminization Syndrome. Report of a Case with Biochemical Psychiatric Studies. *Obstetrics and Gynecology* 26: 499–503.

Terlecki, Melissa S. 1995. The Effects of Long-Term Practice and Training on Mental Rotation. *Dissertation Abstracts International: Section B: The Sciences and Engineering* 65 (10-B): 5434.

Thompson, Eileen G., Irene T. Mann, and Lauren J. Harris. 1981. Relationships among Cognitive Complexity, Sex, and Spatial Task Performance in College Students. *British Journal of Psychology* 72: 249–56.

Tiffin, J. 1948. *Purdue Pegboard*. Chicago: Science Research Associates.

Toniolo, Daniela, and Flavio Rizzolio. 2007. X Chromosome and Ovarian Failure. *Seminars in Reproductive Medicine* 25 (4): 264–71.

Udry, J. R., N. M. Morris, and J. Kovenock. 1995. Androgen Effects on Women's Gendered Behavior. *Journal of Biosocial Science* 27: 359–68.

Valian, Virginia. 2006. Women at the Top in Science—and Elsewhere. In *Why Aren't More Women in Science?* ed. Stephen J. Ceci and Wendy M. Williams, 27–37. Washington, D.C.: American Psychological Association.

Van Goozen, Stephanie H. M., Ditte Slabbekoorn, Louis J. G. Gooren, and Geoff Sanders. 2002. Organizing and Activating Effects of Sex Hormones in Homosexual Transsexuals. *Behavioral Neuroscience* 116: 982–88.

Vandenberg, Steven G., and Allan R. Kuse. 1978. Mental Rotations, a Group Test of Spatial Visualization. *Perceptual and Motor Skills* 47: 599–604.

Vasquez, Melba J. T., and James M. Jones. 2006. Increasing the Number of Psychologists of Color: Public Policy Issues for Affirmative Diversity. *American Psychologist* 61: 132–42.

Vawter, Marquis P., Simon Evans, Prabhakara Choudary, Hiroaki Tomita, Jim Meador-Woodruff, Margherita Molnar, Jun Li, Juan F. Lopez, Rick Myers, David Cox, Stanley J. Watson, Huda Akil, Edward G. Jones, and William E. Bunney. 2004. Gender-Specific Gene Expression in Post-Mortem Human Brain: Localization to Sex Chromosomes. *Neuropharmacology* 29: 373–84.

Voyer, Daniel, Albert Postma, Brandy Brake, and Julianne Imperato-McGinley. 2007. Gender Differences in Object Location Memory: A Meta-Analysis. *Psychonomic Bulletin and Review* 14: 23–38.

Wai, Jonathan, David Lubinski, and Camilla P. Benbow. 2005. Creativity and Occupational Accomplishments among Intellectually Precocious Youths: An Age 13 to Age 33 Longitudinal Study. *Journal of Educational Psychology* 97 (3): 484–92.

Wainer, Howard, and Linda S. Steinberg. 1992. Sex Differences in Performance on the Mathematics Section of the Scholastic Aptitude Test: A Bidirectional Validity Study. *Harvard Educational Review* 62: 323–36.

Wallen, Kim. 2005. Hormonal Influences on Sexually Differentiated Behavior in Nonhuman Primates. *Frontiers in Neuroendocrinology* 26: 7–26.

Watson, Neil V., and Doreen Kimura. 1991. Nontrivial Sex Differences in Throwing and Intercepting: Relation to Psychometrically-Defined Spatial Functions. *Personality and Individual Differences* 12: 375–85.

Webb, Rose Mary, David Lubinski, and Camilla P. Benbow. 2002. Mathematically Facile Adolescents with Math-Science Aspirations: New Perspectives on Their Educational and Vocational Development. *Journal of Educational Psychology* 94 (4): 485–94.

Wennerås, Christine, and Agnes Wold. 1997. Nepotism and Sexism in Peer-Review. *Nature* 387 (6631): 341–43.

Whiting, Beatrice, and Carolyn Edwards. 1973. A Cross-Cultural Analysis of Sex Differences in the Behavior of Children Aged Three through Eleven. *Journal of Social Psychology* 91: 177–88.

Willerman, Lee, Robert Schultz, J. Neil Rutledge, and Erin D. Bigler. 1991. In Vivo Brain Size and Intelligence. *Intelligence* 15: 223–28.

Williams, Christina L., Allison M. Barnett, and Warren H. Meck. 1990. Organizational Effects of Early Gonadal Secretions on Sexual Differentiation in Spatial Memory. *Behavioral Neuroscience* 104: 84–97.

Williams, Christina L., and Warren H. Meck. 1991. The Organizational Effects of Gonadal Steroids on Sexually Dimorphic Spatial Ability. *Psychoneuroendocrinology* 16: 155–76.

Wilson, J. R., and S. G. Vandenberg. 1978. Sex Differences in Cognition: Evidence from the Hawaii Family Study. In *Sex and Behavior*, ed. T. E. McGill, D. A. Dewsbury, and B. D. Sachs, 317–35. New York: Plenum.

Wisniewski, Amy B., Claude J. Migeon, Heino F. Meyer-Bahlburg, John P. Gearhart, Gary D. Berkovitz, Terry R. Brown, and John Money. 2000. Complete Androgen Insensitivity Syndrome: Long-Term Medical, Surgical, and Psychosexual Outcome. *Journal of Clinical Endocrinology and Metabolism* 85 (8): 2664–69.

Witkin, Herman A. 1950. Individual Differences in Ease of Perception of Embedded Figures. *Journal of Personality* 19: 1–15.

———. 1967. A Cognitive Style Approach to Cross-Cultural Research. *International Journal of Psychology* 2: 233–50.

Wittig, Michele A., and Mary J. Allen. 1984. Measurement of Adult Performance on Piaget's Water Horizontality Task. *Intelligence* 8: 305–13.

Woodworth, R. S., and F. L. Wells. 1911. Association Tests. *Psychological Monographs* 13 (57): 1–85.

Xu, Jun, and C. M. Disteche. 2006. Sex Differences in Brain Expression of X- and Y-Linked Genes. *Brain Research* 1126: 50–55.

Conclusion
Why It All Matters and What Is To Be Done

Charles Murray

The essays in this volume have presented spirited defenses of irreconcilable positions. The authors disagree not just on the details, but on the very answer to the question that brought us together: Do men and women have innate cognitive differences that importantly explain their differential representation in the sciences? There is no point in trying to present an above-the-battle, evenhanded assessment of where the authors leave us. Their arguments and evidence are complex and need to be confronted in all their complexity.

Beyond these considerations, my own opinions favor one side of the debate, and it would be disingenuous to pretend otherwise. I believe the debate is eventually going to be resolved in favor of the position represented by Professors Baron-Cohen, Geary, Haier, Kimura, Levy, Sommers, and Wax, and this will happen in a matter of years, not decades. This does not mean that the material presented by Professors Aronson, Barnett, Ellison, Sabattini, and Spelke is "wrong" in its specifics. I accept the data they presented at face value. Rather, the study of sex differences has moved too far beyond the phenotype for data based on the phenotype to be decisive. In this concluding chapter, I will first present my reasons for taking that position, and then turn to why it all matters, and what is to be done.

In focusing on the dynamic of the debate, I am influenced by Thomas Kuhn's description of scientific revolutions.[1] The mainstream paradigm from the late 1960s into the new century was social construction based on the equality premise: the premise that men and women have no important cognitive differences. I use *social construction* to mean the argument that society has invented gender roles that are not grounded in real sex differences. I use *cognitive* in its broad sense, including the abilities described by Howard Gardner as intrapersonal intelligence and interpersonal intelligence, as well as intellectual ability ordinarily defined.[2]

Social construction couldn't survive as the paradigm without consensus acceptance of the equality premise among social scientists. The premise never had that kind of support among biologists, but for a long time they didn't count. When it came to thinking about sex differences during social construction's zenith, only a handful of the boldest social scientists in academia would say publicly that men and women might be cognitively different for genetic reasons. The experience of one of the few who did, sociologist Steven Goldberg, is illustrative. His work, *The Inevitability of Patriarchy*, published in 1973, is listed in *The Guinness Book of World Records* as the book rejected by the most publishers (sixty-nine rejections by fifty-five publishers) before finally being accepted.[3]

But starting in the 1970s and accelerating during the 1980s, cracks in the paradigm appeared and spread. The first pivotal event was the publication of E. O. Wilson's *Sociobiology* in 1975—not because of what it said about sex differences specifically, but because it provided a beachhead for the study of human nature through the instruments of biology, not sociology.[4] The subsequent scientific work proceeded discreetly, almost furtively, through articles published in technical journals written in academic prose that would not catch the attention of the media. But the work accumulated. When David Geary published *Male, Female* in 1998—another pivotal event—the technical literature had already become so extensive that Geary's bibliography was fifty-three pages long.[5]

The completion of the first mapping of the human genome in early 2001 marked the third pivotal event. Substantive findings from that accomplishment were still a few years away, but everyone knew that the floodgates were about to open. The dynamic of the dialogue changed accordingly, and the prevailing paradigm became fragile. As of 2009, we already live in a different world. I realize that defenders of the equality premise still managed to have Larry Summers fired from the presidency of Harvard University just three years ago, but that had the flavor of the Catholic Church's trial of Galileo—the response of an orthodoxy that knows its defenses have been breached.

Here are four specific ways in which I believe the direction and momentum of the debate have moved beyond recall:

- Biology is irreversibly displacing social construction as the paradigm for investigating the phenotype.

- The scientific evidence for significant, genetically grounded, cognitive sex differences is already too strong to be denied.

- The differences already known to exist must differentially affect the distribution of occupational preferences among men and women.

- The differences already known to exist give men three genetic advantages that will preserve their disproportionate contributions at the extremes of scientific accomplishment until genetic engineering alters them.

Here are my reasons for thinking that all four statements, so inflammatory that they would be shouted down if said aloud at any number of academic conventions, are actually not empirically controversial.

Biology is irreversibly displacing social construction as the paradigm for investigating the phenotype. I am premature in making that claim if the measure is based on classes taught in academia. The catalogs are still full of postmodernist courses that explain all cognitive sex differences as cultural artifacts. But if the measure is the body of new work being produced and the degree to which it forms the basis of the conversation, biology—embracing genetics, neuroscience, and evolutionary psychology—is in and social construction is out. One symptom is the reception of Steven Pinker's *The Blank Slate* in 2002. A bestseller and a finalist for the Pulitzer Prize, its enthusiastic reception would have been unthinkable a decade earlier. Pinker himself was aware of the shift, opening the book by drawing attention to the discrepancy between the reaction of his colleagues as he wrote it—"Are there really people out there who still believe that the mind is a blank slate?"—and the vitriolic response to books such as *The Nurture Assumption*, *A Natural History of Rape*, and *The Bell Curve* during the 1990s.[6] A more recent symptom is the supplemental issue of the *American Journal of Sociology* published in 2008 devoted to the topic, "Exploring Genetics and Social Structure." The subtext threaded throughout the issue is that sociology is in danger of rendering itself obsolete unless it incorporates the role of genes.[7] The fulfillment of E. O. Wilson's prediction in *Consilience* that the twenty-first century would see the integration of the social and biological sciences is starting to unfold.[8]

Once it is taken for granted that human nature is not a blank slate and that it is shaped by evolutionary pressures, then it must also be taken for granted that males and females are cognitively different. This implication is why the blank-slate view of human nature survived so tenaciously in discussions of sex differences in the social sciences long after no one in the hard sciences believed it. The different evolutionary pressures facing males and females have always been obvious. To admit a genetically grounded human nature is to admit defeat.

To see why this must be so, imagine a situation in which you are behind a veil of ignorance about the differing positions of men and women in today's society and have never heard of evolutionary psychology. You are aware only of the most self-evident physiological differences between men and women, and you accept that natural selection has governed human evolution. As this naïve layperson, you are asked to think about why women evolved to be (apparently, from everyday observation) so much more emotionally wrapped up in their children from day to day than men are.

You can answer the question even if the only thing you know is that females are able to nurse infants while males cannot, and that breast-feeding is by far the most effective way to keep infants alive in preindustrial societies. From this simplest of all facts about sex differences and the survival of infants, it is easy to understand why females were always the sex that took primary care of young children. Then think about reproductive fitness. What kind of women were most likely to perpetuate their genes? Answer: Women who kept their infants and toddlers alive long enough to reproduce—which means that natural selection would favor women who were so devoted to their offspring that they did, in fact, nurse them faithfully, as well as lavish all the other time and effort that goes into keeping an infant or toddler alive. Women who were emotionally indifferent to infants had a lower probability of keeping their children alive. Men who were emotionally indifferent to infants did not suffer nearly the same fitness penalty. They had to be able to get sexual access to women, but their reproductive fitness did not depend on being devoted to their offspring from moment to moment.

The full explanations for the differentials in male and female parental investment are much more sophisticated than this, as the presentations of Levy and Kimura and Geary indicate.[9] I am offering a simple "of course" proposition. *Of course* women are genetically hardwired to be more nurturing of small children and more psychologically absorbed in their welfare than men are. How can anyone doubt it? From that one evolutionarily inevitable difference, a wide variety of cognitive implications follow. It is not possible to accept the basics of evolutionary theory without expecting major cognitive differences in males and females. That reality is increasingly recognized in the social sciences, and the spread of that recognition cannot be stopped.

The scientific evidence for significant, genetically grounded, cognitive differences is already too strong to be denied. If one sets out to prove the hypothesis that black swans exist, it is not necessary to find flocks of them. One is sufficient. As of 2009, many neural black swans have been spotted to prove the hypothesis that men and women differ cognitively. Richard Haier's report on the different regions of the brain used by men and women for comparable tasks and the different roles played by grey matter and white matter for men and women falls into this category. The meaning of these data still needs to be elaborated, but that sex differences exist already seems well established.

Simon Baron-Cohen has found a hormonal black swan. In his presentation, Baron-Cohen focused on observed differences in males and females, mentioning only in passing his extensive research on the relationship of fetal testosterone (which varies systematically by sex) to the development of the brain and subsequent differences in the behavior of children. I urge readers to explore the extensive evidence for this brief allusion.[10] The ultimate test of a scientific theory is its ability to predict. The evidence is compelling that variations in fetal testosterone predict a set of behavioral results in infants, toddlers, and five-year-olds—within sexes as well as between sexes—in ways that accord with Baron-Cohen's theoretical understanding of male–female differences.[11]

Other black swans are documented in the Levy and Kimura presentation.[12] Taken as a body, the existing findings already prove significant differences in the ways that male and female brains are organized, the ways that they work, and even the sizes of their different portions.[13] Nothing is going to make those data go away. The binary yes/no question has been answered. The rest is a matter of elaboration.

The differences already known to exist must differentially affect the distribution of occupational preferences among men and women. The existence of these innate cognitive differences is not important for the vast majority of occupations. For all but a small minority of jobs, such differences give employers no valid reason to prefer one sex for employment or promotions. That statement certainly applies to jobs in math and the sciences. More broadly, there is a longstanding consensus among psychometricians that men and women have the same mean full-scale IQ.[14] There is no "smarter sex" in any global sense of that term, just areas of comparative advantage for one sex or the other in specific subtests. These differences may be statistically significant, but they are substantively trivial in most workplaces.

The powerful effects of the innate differences in day-to-day life involve career preferences. Little boys and little girls have different play preferences. The presentations of Geary, Levy and Kimura, and Baron-Cohen cite some of the scholarly evidence that these different preferences are systematic and pervasive, and appear very early—in some cases, in the first day of life. The sources they cite are based on scientific observation and valid samples, not parental anecdotes. But I cannot resist noting that much of the erosion of belief in the equality premise among intellectuals during the 1980s and 1990s was probably fostered by their experiences when they tried to raise their own children in gender-neutral ways. Given dolls, their little boys used them to pound things. Given trucks, their little girls used them to transport make-believe groups of friends to parties. The impossibility of suppressing such sex differences has become undeniable to many parents who were once sure that they were the products of socialization.

The relationships between play preferences as children and career preferences as adults hang together. It makes sense that girls who prefer dolls and social play are more drawn to intellectual challenges involving people than to intellectual challenges involving the design of machines and bridges, more drawn to professions that call for close interactions with people than professions that put them alone at a work station in a laboratory. Among the subset of women who are attracted to the sciences, those same differences come into play. They are reflected in the greater likelihood that a woman will become a practicing physician than a medical researcher, and that she is more likely to major in biology or anthropology than electrical engineering. It's not a matter of men being smarter than women. It's a matter of what pursuits give them more satisfaction and enjoyment.

The differences already known to exist give men three genetic advantages that will preserve their disproportionate contributions at the extremes of scientific accomplishment until genetic engineering alters them. Achievement at the extremes in math and science is different from merely having a successful career as a researcher or teacher. I am referring to achievements that win Nobel Prizes or Field Prizes or are included in histories of science.

Here I will ignore the case for a genetic male advantage in visuospatial skills and abstract thinking that several of the other essays discussed. I consider it to be an overpoweringly strong case, but the Spelke and Ellison presentation disputes it with data, so I put it aside, focusing instead on three other characteristics that, statistically, must give males a genetic advantage in scientific achievement at the highest levels even if males do *not* have an advantage in visuospatial skills and abstract thinking.

The first facilitator is freakishly high intellectual ability for the scientific discipline in question. Math and the sciences are unusual in this regard. The most successful politicians, business executives, journalists, or movie directors are almost universally much smarter than average, but not necessarily at the very top of the distribution.[15] In contrast, the greatest mathematicians and scientists—especially scientists in the most mathematically demanding disciplines, such as physics—do tend to be freakishly smart in a psychometric sense, meaning three or four standard deviations above the mean—so high that the top scores on ordinary tests of academic ability such as the SAT do not identify them.[16] Here, the greater variance in IQ among males than females (along with greater variance in other characteristics) referenced by several of the other essays comes into play. If there are several times as many men as women with mathematical ability at the freakish extreme, then a large majority of great mathematicians will always be male.

The second facilitator is undistracted concentration on the work. Great achievement in almost every field is associated with a crushing workload pursued with single-minded intensity. It is one of the most universal findings in the literature about extraordinary success among creative people.[17] One thread of this literature estimates the amount of knowledge required to achieve expertise—fifty thousand "chunks" in Herbert Simon's calculation, accumulated over about ten years of experience.[18] Another explores the role of thousands of hours of practice, study, and labor after the expertise is achieved; the great accomplishments of the famous scientists and mathematicians typically represent only the visible tip of their iceberg of effort.[19]

This characteristic of great achievement gives men two genetic advantages. First, men are more competitive and aggressive than women.[20] Once again, there are known physiological explanations, with testosterone at the top of the list. There are downsides to this sex difference, reflected, for example, in statistics showing that men throughout the world commit the overwhelming proportion of violent crime.[21] But high levels of competitiveness and aggressiveness facilitate obsessive effort toward an objective.

The genetic advantage in competitiveness is probably less important than the more obvious genetic difference that I discussed earlier: Men have not been subject to the same evolutionary selection as women to be nurturing parents. Women are more attracted to children than are men, respond to them more intensely on an emotional level, and get more and different kinds of satisfactions from nurturing them.[22] Many of these behavioral differences have been linked with biochemical differences between men and women, as some of the presentations have emphasized.

This consideration comes into play because of another consistent finding in the study of great achievement: The peak of productivity is reached around the age of forty, following years of intense apprenticeship that have gradually morphed into the realized capabilities that enable great achievement.[23] These are precisely the years during which most women must bear children if they are to bear them at all. Women who spend a substantial number of years from twenty to forty focused on anything besides their math or science are just as unlikely to produce the greatest work as men with the same ability would be if they divided their energies. But in the case of women, it is not just the occasional eccentric among the exceptionally gifted who divides her energies. A majority do, for a reason that is understandable and laudable, and is not going to go away: they want to be mothers.

Women who have children but continue their jobs full-time are still not in the same position as men if the goal is exceptional scientific achievement.

If we are talking about an ordinarily successful career, children need not be a barrier, as millions of mothers with daycare, nannies, or stay-at-home dads demonstrate every day. The problem comes when we move to the possibilities for the highest-level accomplishment in the sciences. For women, the distractions of parenthood are greater than for men. To put it in a way that most readers with children will recognize, the father of a child who has the flu can go to work and not give it a thought throughout the entire day. Hardly any mother can help having it on her mind throughout the day, no matter how good her daycare arrangement or nanny might be. For achieving at the highest levels, worrying about the children disadvantages women.

To avoid misunderstanding, let me acknowledge explicitly that none of these generalizations applies to all women. Do some women have the stratospherically high intellectual ability to be great mathematicians? Yes. Do some women forgo motherhood? Yes. Are some mothers able to do their scientific work undistracted by thoughts of their children? Yes. I am making a set of probabilistic statements. If more men than women have the requisite intellectual ability, then, *ceteris paribus*, more men will achieve at the highest levels. If, among those who do have the requisite intellectual ability, more women than men spend a large part of the crucial years from twenty to forty doing something besides the apprenticeship, then, *ceteris paribus*, more men will achieve at the highest levels. If, among those who do have the requisite intellectual ability and have completed the apprenticeship, more men than women can work brutally long hours and do so undistracted by thoughts about children, then, *ceteris paribus*, more men than women will achieve at the highest levels.

◆ ◆ ◆

Why does it all matter? In important and positive ways, it is mattering less and less. The world for talented young women in science and math is incomparably more welcoming in 2009 than it was fifty years ago. A personal recollection is relevant: My eldest sister set out to major in architecture at Iowa State University in 1955. When she was an eighteen-year-old freshman, in one of her first days of college classes, a professor looked down at her—the only girl in the room—and told the class he didn't think girls belonged in engineering courses. The effect of that kind of hostility on young women with an interest in science is hard to exaggerate.

The world really is better now. When my sister heard those discouraging words in the mid-1950s, women obtained just 0.4 percent of the engineering bachelor's degrees awarded each year. In 2005, 18 percent of engineering degrees went to women. The increases in the proportions of engineering master's degrees and doctorates going to women were even larger.[24] The same thing happened in other technical fields. In the physical sciences, for example, women got 12 percent of the bachelor's degrees and 3 percent of the doctorates in 1960. In 2005, they got 42 percent of the bachelor's degrees and 30 percent of the doctorates. In math and statistics, women got 27 percent of the bachelor's degrees and 6 percent of the doctorates in 1960. In 2005, the comparable figures were 45 percent and 30 percent. These dramatic increases do not mean that bias against women is no longer an issue. But by any quantitative indicator, young women who show scientific or math talent today are getting much more encouragement than they used to. That's good.

But I must also offer a warning. Earlier, I argued that from the late 1960s through the end of the century, the academic mainstream in the social sciences embraced the equality premise. But so did the political mainstream. Virtually every social policy initiated since the late 1960s has reflected the assumption that all groups of people are cognitively indistinguishable. Since we observe very large group differences in the phenotype, the equality premise forces the conclusion that when we see inequalities, the only cause must be environmental disadvantages afflicting the group with the lower income, education, or social status. Everything that we associate with the phrase "politically correct" eventually comes back to the equality premise. In social policy, the statistical tests for uncovering job discrimination are based on the equality premise. Affirmative action in all its forms assumes there are no innate cognitive differences between any of the groups it seeks to help and everyone else.

These academic and political mainstreams have not been limited to one part of the ideological spectrum. I can attest from my experience with the reaction to *The Bell Curve* that conservatives and libertarians are as uncomfortable with the idea of innate cognitive group differences as liberals are. Hence a question that we need to start thinking about: What will happen when it is proved beyond a shadow of a scientific doubt that important genetic group differences do exist? I urge that we start thinking about the answer, because, in my judgment, that hypothetical situation will soon be upon us for real. Perhaps we have as much as a decade before it happens, but it cannot be much more than that and may be less. It will probably come first with regard to differences between males and females—partly because the science is already more advanced, and partly because the study of sex differences remains less taboo than the study of race and ethnic differences.

This new knowledge need not be scary. After the puzzle has been pieced together, genetic group differences will still be a matter of probabilities, in which two groups have statistically significant distributions but overlap substantially. Genetic group differences will cut both ways, with every group having its own strengths and attractions. I doubt that many women will wish they were men (or vice versa) when important innate sex differences have been proved beyond doubt. As the new knowledge extends to other areas, I doubt that many people will wish that they had been born Chinese instead of Swedish (or vice versa), white instead of black (or vice versa), or inclined to be an English professor instead of an engineer (or vice versa), no matter what science may eventually tell us about the ways that these groups are genetically distinct. Human beings have a marvelous capacity to observe group differences and come up with a calculus that makes their own group preferable in their own eyes.

The ominous implications do not derive from the science itself, but from the overreaction that may follow after decades in which the existence of group differences has been so passionately denied. A process of cognitive dissonance is underway in which everyone knows—even those who are trying hardest to maintain the faith—that some taboo ideas are likely to be true. When cognitive dissonance is resolved, it seldom takes the form of "never mind; I guess I was wrong about that." Too often, it produces the fervor of the newly converted. I cannot be precise about predictions, because we have never seen anything quite like the situation that is playing out. The only thing I am sure of is that a great deal of emotional and intellectual energy has been invested in the proposition that different outcomes for different groups are produced by injustices that can be fixed with the right policies. I am forecasting a very tough situation facing the people who have been committed to that proposition. To put it in terms of the topic of this collection of essays, I am forecasting that they will have to acknowledge that women will always be a minority on science faculties and will always win fewer Field Prizes and Nobel Prizes for physics than men do—even in a perfectly fair world.

Acknowledging such realities will not mean that political liberals must also give up on liberal causes. On the contrary, the proof of innate aspects of human capital can logically be employed as a justification for greater governmental redistribution to achieve equality of material outcomes to make up for the inequalities of human capital that nature has imposed on us. Similarly, acknowledging such realities confers no necessary benefit on political conservatives. Inequalities of opportunity will remain, and the crusades for equal opportunity that antedated the equality premise—the suffrage movement in the early part of the twentieth century or the civil rights movement of the 1950s and early 1960s—were led by liberals, not conservatives.

The dangers attending the coming demise of the equality premise affect us all. To ward them off, all of us, from the hard and soft sciences, from the political left and right, male and female, need to embrace once again the old American ideal of treating people fairly as individuals. Steven Pinker put that ideal in today's language in The Blank Slate, writing that "equality is not the empirical claim that all groups of humans are interchangeable; it is the moral principle that individuals should not be judged or constrained by the average properties of their group."[25] That principle fell on hard times with the advent of the equality premise. Reanimating it as the central moral imperative of the American tradition is our way to deal with the new scientific knowledge that is rushing in upon us.

Notes

1. Kuhn 1962.
2. Gardner 1985, chapter 10.
3. Goldberg 1973, 2.
4. Wilson 1975.
5. Geary 1998.
6. Pinker 2002, vii–viii.
7. For the table of contents of the special issue, see *American Journal of Sociology* 2008.
8. Wilson 1998.
9. Geary 1998. See also Jones et al. 2003; Kimura 1999, 11–30; and Baron-Cohen 2003, chapter 9.
10. Baron-Cohen started from Norman Geschwind's hypothesis that fetal testosterone affects the growth rate of the hemispheres of the brain, with higher testosterone promoting faster growth of the right hemisphere and inhibiting growth of the left hemisphere, with corresponding implications for the behavior and mental characteristics of children. Baron-Cohen and his colleagues were able to analyze the amniotic fluid for a substantial number of pregnancies, measure the level of fetal testosterone, and then test the children produced by those pregnancies. Measures of social behavior and cognitive functioning correlated with level of fetal testosterone exactly as predicted by Geschwind's hypothesis. For a nontechnical presentation, see Baron-Cohen 2003, chapter 8. For a more detailed technical presentation, see Chapman et al. 2006 and Auyeung et al., forthcoming.
11. Baron-Cohen 2003, chapter 8.
12. For a complementary overview, see Cahill 2005.
13. See Goldstein et al. 2001. Since many readers have read Gould's *Mismeasure of Man* (1981) and are under the impression that variation in brain volume among humans is not related to cognitive functioning, I should note that magnetic resonance imaging (MRI) studies of brain size have ended the uncertainty about the existence of its relationship with IQ. Meta-analyses of MRI and other *in vivo* studies indicate that the correlation between brain size and IQ is about .40 (Jensen 1998, 147).
14. For a review of the evidence that male and female IQ is the same, see Jensen 1998, 536–42. The underlying problem is that the subtests in IQ tests have been developed and normed in ways that tend to push male and female IQs toward the same mean IQ (for example, items that show a large sex difference are usually discarded). For the controversial arguments that men have a higher mean IQ than women, see Ankney 1992; Lynn 1999; and Lynn and Irwing 2004.
15. Gottfredson 1997; Herrnstein and Murray 1994, chapters 2 and 3.
16. The IQs of important intellectuals of the past have been retrospectively estimated using behavioral correlates of IQ (for example, the age at which a child begins to read). The evidence of extraordinary precocity, denoting IQs at least three or four

standard deviations above the mean, is common among great scientists and close to universal among great mathematicians. The most ambitious example of this analysis is Cox 1926. For the quantitative relationship of similarly high IQ, measured using the Study of Mathematically Precocious Youth, to outcome measures such as number of journal articles and patents, see Park et al. 2008.

17. For a summary of the literature and reference to other sources, see Simonton 1988, chapter 4.

18. Simon 1972.

19. For the role of practice, see Ericcson and Tesch-Romer 1993.

20. Byrnes et al. 1999; Dabbs and Dabbs 2000.

21. Wilson and Herrnstein 1985, chapter 4.

22. Baron-Cohen 2003 has the most integrated theory about differences between males and females, with the nurturing aspects subsumed under the larger argument. For an analysis specifically devoted to sex differences in nurturing, written by a committed feminist who is also a scientist (an anthropologist), see Hrdy 1999. For a short review of studies on the importance of children and of the biological sources of nurturing differences, see Rhoads 2004, 190–222.

23. Simonton 1984, chapter 6.

24. The data in this paragraph are taken from U.S. Department of Education, Institute of Education Sciences, National Center for Education Statistics 2008, tables 294, 300, and 301.

25. Pinker 2002, 340.

References

American Journal of Sociology. 2008. Table of contents of special issue, Exploring Genes and Social Structure. *American Journal of Sociology*. 114 (supplement). http://www.journals.uchicago.edu/toc/ajs/2008/114/2 (accessed January 21, 2009).

Ankney, Davison. 1992. Sex Differences in Relative Brain Size: The Mismeasure of Woman, Too? *Intelligence* 16 (3-4): 329–36.

Auyeung, Bonnie, Simon Baron-Cohen, Emma Ashwin, Rebecca Knickmeyer, Kevin Taylor, and Gerald Hackett. Forthcoming. Fetal Testosterone and Autistic Traits. *British Journal of Psychology.*

Baron-Cohen, Simon. 2003. *The Essential Difference: Male and Female Brains and the Truth about Autism.* New York: Basic Books.

Byrnes, James P., David C. Miller, and William D. Schafer. 1999. Gender Differences in Risk Taking: A Meta-Analysis. *Psychological Bulletin* 125 (3): 367–383.

Cahill, Larry. 2005. His Brain, Her Brain. *Scientific American*. April. http://www.sciam.com/article.cfm?id=his-brain-her-brain (accessed January 19, 2009)

Chapman, Emma, Simon Baron-Cohen, Bonnie Auyeung, Rebecca Knickmeyer, Kevin Taylor, and Gerald Hackett. 2006. Fetal Testosterone and Empathy: Evidence from the Empathy Quotient (EQ) and "Reading the Mind in the Eyes" Test. *Social Neuroscience* 1 (2): 135–48.

Cox, C. M. 1926. *The Early Mental Traits of Three Hundred Geniuses.* Stanford, Calif.: Stanford University Press.

Dabbs, J. M., and M. G. Dabbs. 2000. *Heroes, Rogues, and Lovers: Testosterone and Behavior.* New York: McGraw-Hill.

Ericcson, A. R. T., and C. Tesch-Romer. 1993. The Role of Deliberate Practice in the Acquisition of Expert Performance. *Psychological Review* 100 (3): 363–406.

Gardner, Howard. 1985. *Frames of Mind: The Theory of Multiple Intelligences.* New York: Basic Books.

Geary, David C. 1998. *Male, Female: The Evolution of Human Sex Differences.* Washington, D.C.: American Psychological Association.

Goldberg, Steven. 1973. *The Inevitability of Patriarchy.* New York: Morrow.

———. 1993. *Why Men Rule: A Theory of Male Dominance.* Chicago: Open Court.

Goldstein, Jill M., L. J. Seidman, Nicholas J. Horton, N. Makris, D. Kennedy, V. Caviness, S. V. Faraone, and M. T. Tsuang. 2001. Normal Sexual Dimorphism of the Adult Human Brain Assessed by *In Vivo* Magnetic Resonance Imaging. *Cerebral Cortex* 11: 490–97.

Gottfredson, Linda S. 1997. Why g Matters: The Complexity of Everyday Life. *Intelligence* 24 (1): 79–132.

Gould, Stephen J. 1981. *The Mismeasure of Man.* New York: W. W. Norton.

Harris, J. R. 1998. *The Nurture Assumption: Why Children Turn Out the Way They Do.* New York: Free Press.

Herrnstein, Richard J., and Charles Murray. 1994. *The Bell Curve: Intelligence and Class Structure in American Life.* New York: Free Press.

Hrdy, Sara Blaffer. 1999. *Mother Nature: A History of Mothers, Infants, and Natural Selection*. New York: Pantheon Books.

Jensen, Arthur R. 1998. The g Factor: *The Science of Mental Ability*. Westport, Conn.: Praeger.

Jones, C. M., V. A. Braithwaite, and S. D. Healy. 2003. The Evolution of Sex Differences in Spatial Ability. *Behavioral Neuroscience* 117: 403–11.

Kimura, Doreen. 1999. *Sex and Cognition*. Cambridge, Mass.: MIT Press.

Kuhn, Thomas S. 1962. *The Structure of Scientific Revolutions*. 2d ed. Chicago: University of Chicago Press.

Lynn, Richard. 1999. Sex Differences in Intelligence and Brain Size: A Developmental Theory. *Intelligence* 27 (1): 1–12.

Lynn, Richard, and Paul Irwing. 2004. Sex Differences on the Progressive Matrices: A Meta-Analysis. *Intelligence* 32: 481–98.

Park, G., D. Lubinski, and C. P. Benbow. 2008. Ability Differences among People Who Have Commensurate Degrees Matter for Scientific Creativity. *Psychological Science* 19: 957–61.

Pinker, Steven. 2002. *The Blank Slate: The Modern Denial of Human Nature*. New York: Viking Penguin.

Rhoads, Steven E. 2004. *Taking Sex Differences Seriously*. San Francisco: Encounter Books.

Simon, Herbert A. 1972. Productivity among American Psychologists: An Explanation. *American Psychologist* 9: 804–5.

Simonton, Dean K. 1984. *Genius, Creativity, and Leadership*. Cambridge, Mass.: Harvard University Press.

———. 1988. *Scientific Genius: A Psychology of Science*. Cambridge: Cambridge University Press.

Thornhill, R., and Palmer, C. T. 2000. *A Natural History of Rape: Biological Bases of Sexual Coercion*. Cambridge, Mass.: MIT Press.

U.S. Department of Education. Institute of Education Sciences. National Center for Education Statistics. 2008. *The Digest of Education Statistics 2007*. By Thomas D. Snyder, Sally A. Dillow, and Charlene M. Hoffman. NCES 2008-022. http://nces.ed.gov/pubs2008/2008022.pdf (accessed January 19, 2009).

Wilson, Edward O. 1975. *Sociobiology*. Cambridge, Mass.: Harvard University Press.

———. 1998. *Consilience: The Unity of Knowledge*. New York: Alfred A. Knopf.

Wilson, James Q., and Richard J. Herrnstein. 1985. *Crime and Human Nature*. New York: Simon and Schuster.

Index

About the Authors

Joshua Aronson is an associate professor of applied psychology at New York University. Through his research on social and psychological influences on motivation and confidence, he seeks to understand and remediate race- and gender-based gaps in educational achievement and standardized test performance. In 1995, Mr. Aronson and his colleague Claude Steele published groundbreaking laboratory studies on "stereotype threat," which they describe as a performance-inhibiting phenomenon that occurs when students face negative stereotypes assigned to their particular race or gender. Since then, Mr. Aronson has published many chapters and scholarly articles on the topic, as well as the guidebook *Improving Academic Achievement: Impact of Psychological Factors on Education*. Mr. Aronson's forthcoming book is titled *The Nurture of Intelligence*. His work has been cited in two recent Supreme Court cases, including *Grutter v. Bollinger*, and it is frequently referred to in policy debates about educational equality. Mr. Aronson has received a number of awards and grants for his research, including Early Career awards from the Society for the Psychological Study of Social Issues and the National Science Foundation and the G. Stanley Hall Award from the American Psychological Association.

Rosalind Chait Barnett is a senior scientist at the Women's Studies Research Center at Brandeis University. She has authored or coauthored over 110 articles and six books, the most recent of which is *Same Difference: How Gender Myths are Hurting Our Relationships, Our Children and our Jobs*, coauthored with Caryl Rivers. She and Ms. Rivers are currently working on a new book, *The Truth about Boys and Girls* (Columbia University Press). Her writing has appeared in many scholarly journals, as well as the *Washington Post*, the *Boston Globe*, the *New York Times Magazine*, the *Los Angeles Times*, and *Self* magazine. Ms. Barnett's current research projects are an examination of the role communities play in the lives of families with school-aged children and an assessment of concerns employees have about adult relatives and elder care and how they relate to employees' well-being and job productivity. Ms. Barnett is the recipient of several national and international awards, including the American Personnel and Guidance Association's Annual Award for Outstanding Research, the Radcliffe College Graduate Society's Distinguished Achievement Medal, and the Harvard University Kennedy School of Government's 1999 Goldsmith Research Award.

Simon Baron-Cohen is a professor of developmental psychopathology at the University of Cambridge, a fellow at Trinity College, Cambridge, and director of the Autism Research Centre at Cambridge. He is best known for his research on autism, particularly his theory that autism is an extreme form of the "male brain," a theory that has led to entirely new ways of comprehending psychological gender differences with regard to empathy and systemizing. Mr. Baron-Cohen has written and edited several books, including *Mindblindness, The Essential Difference: The Truth about the Male and Female Brain*, and, most recently, *Prenatal Testosterone in Mind.* His work has had significant influence in the fields of developmental and clinical psychology, psychiatry, cognitive neuroscience, primatology, and philosophy of mind. Mr. Baron-Cohen is president of the Psychology Section of the British Association for the Advancement of Science and vice president of the National Autistic Society. Last year, he received the President's Award for Distinguished Contributions to Psychological Knowledge from the British Psychological Society.

Katherine Ellison is a graduate student in clinical psychology at Fairleigh Dickinson University in Madison, New Jersey.

David C. Geary is the Curators' Professor and former chair of the Department of Psychological Sciences at the University of Missouri. He specializes in cognitive developmental psychology with an interest in mathematical learning and in evolution. Mr. Geary has published more than 150 articles and chapters across a wide range of topics, as well as three books, the most recent of which is *The Origin of Mind: Evolution of Brain, Cognition, and General Intelligence*, published in 2005. His 1998 book, *Male, Female: The Evolution of Human Sex Differences*, explores many frequently asked questions about gender differences, such as why men tend to be more aggressive than women. A member of the President's National Mathematics Panel, Mr. Geary is the lead investigator for a longitudinal study of children's mathematical development and learning disabilities, and he was a key contributor to the 1999 Mathematics Framework for California public schools. Among many other distinctions, he received the 1996 Chancellor's Award for Outstanding Research and Creative Activity in the Social and Behavioral Sciences and a scientific MERIT award from the National Institutes of Health.

Richard Haier is professor-in-residence (emeritus) in the Pediatric Neurology Division of the School of Medicine at the University of California, Irvine. He completed his graduate education in Psychology at the Johns Hopkins University and has held positions at the National Institute of Mental Health and the Brown University School of Medicine. Mr. Haier is best known for his research using neuroimaging to study brain structure and function as they relate to human intelligence. Mr. Haier has published over 125 papers and book chapters on individual differences and higher cognitive functions in leading scientific journals, including *Intelligence*, *NeuroImage*, and the *Proceedings of the National Academy of Sciences*.

Doreen Kimura has held a post-retirement visiting professorship at Simon Fraser University in British Columbia since 1998. Her undergraduate and graduate education was completed at McGill University in Montreal. She subsequently held postdoctoral fellowships at UCLA Medical Center, Kantonsspital Zürich, and College of Medicine, McMaster University, Hamilton, Canada. She then spent thirty-one years at the University of Western Ontario, London, in Psychology and Neuroscience. Her research has focused on neural mechanisms of cognitive function in humans. This has included outlining the differing functions of the left and right hemispheres of the brain, the relation between control of motor and communicative functions in the left hemisphere, individual differences in brain organization with special reference to sex and handedness differences, and the influence of sex hormones on cognitive pattern. Special awards include distinguished science awards from the Canadian Psychological Association and the Canadian Society for Brain, Behaviour and Cognitive Science, and the Kistler medal from the Foundation for the Future.

She is a member of the Royal Society of Canada and has received honorary Doctor of Laws degrees from Canadian universities.

Jerre Levy completed her undergraduate studies in experimental psychology at the University of Miami in 1962. Her doctoral research in psychobiology was conducted at the California Institute of Technology, from which she received her Ph.D. in 1970. She held postdoctoral fellowships in biopsychology (University of Colorado, 1970–71) and molecular genetics (Oregon State University, 1971–72) before joining the faculty of the Department of Psychology at the University of Pennsylvania (Assistant Professor, 1972–75; Associate Professor, 1975–77). In 1977, she moved to the University of Chicago, where she was on the faculty of the Department of Psychology (formerly, Behavioral Sciences), the College, and the Committee on Neurobiology (Associate Professor, 1977–83; Professor, 1983–2000; Professor Emerita, 2000–present). Her research has been in the area of cognitive neuroscience, with a focus on the asymmetric functions of the normal brain, communication between the two cerebral hemispheres, and individual differences in brain organization and behavior.

Charles Murray, a social scientist and writer, is the W. H. Brady Scholar in Culture and Freedom at the American Enterprise Institute. He is the author of *Losing Ground: American Social Policy, 1950-1980* (1984), *In Pursuit: Of Happiness and Good Government* (1988), *The Bell Curve: Intelligence and Class Structure in American Life* (with Richard J. Herrnstein, 1994), *What It Means to Be a Libertarian: A Personal Interpretation* (1997), *Human Accomplishment: The Pursuit of Excellence in the Arts and Sciences, 800 B.C. to 1950* (2003), *In Our Hands: A Plan to Replace the Welfare State* (2006), and *Real Education: Four Simple Truths for Bringing America's Schools Back to Reality* (2008). He is the recipient of AEI's 2009 Irving Kristol Award for extraordinary contributions to improved public policy and social welfare.

Laura Sabattini has extensive expertise on work-life quality issues, organizational effectiveness, and gender-based stereotyping. She leads research related to women's leadership, talent management strategies, and barriers to women's corporate advancement and is the co-leader of the Work-Life Issue Specialty Team at Catalyst. Dr. Sabattini was formerly an adjunct faculty member at the

University of California at Santa Cruz, where she taught classes in social and organizational psychology and research methodology. She also conducted and published research investigating the challenges that women face in the workplace, the division of household labor between women and men, and the strategies women and men use to manage work and family commitments. She received her B.A. in organizational and work psychology from the University of Padua, Italy, and her M.S. and Ph.D. in social psychology from the University of California at Santa Cruz.

Christina Hoff Sommers is a resident scholar at AEI. She was a professor of philosophy at Clark University from 1981 to 1996. Ms. Sommers specializes in ethics and contemporary moral theory and has published many scholarly articles in such journals as the *Journal of Philosophy* and the *New England Journal of Medicine*. She edited *Vice and Virtue in Everyday Life*, one of the most popular ethics textbooks in the country. Ms. Sommers became known to the wider public as the author of *Who Stole Feminism? How Women Have Betrayed Women*. Her book *The War Against Boys* received widespread attention and praise and was excerpted for a cover story in the *Atlantic Monthly*. It was included in the *New York Times'* "Notable Books of the Year." Her most recent book, *One Nation Under Therapy*, coauthored with Sally Satel, M.D., has received a great deal of attention and critical acclaim. Ms. Sommers's articles have appeared in the *Wall Street Journal*, the *New York Times*, the *Washington Post*, the *Boston Globe*, *USA Today*, *Weekly Standard*, *National Review*, the *New Republic*, and the *American*.

Elizabeth S. Spelke is the Marshall L. Berkman Professor of Psychology at Harvard University. She directs Harvard's Laboratory for Developmental Studies and co-directs its Mind, Brain, and Behavior Initiative. Ms. Spelke's research on human cognition and developmental psychology has been published in numerous journals and publications, including the *American Psychologist*, *Nature*, *Science*, *Cognition*, and the *Quarterly Journal of Experimental Psychology*. A distinguished leader in the field, Ms. Spelke received the Distinguished Scientific Contribution Award from the American Psychological Association and the William James Award from the American Psychological Society in 2000. The following year, she was named by *Time* magazine as one of America's best in science and medicine.

Amy L. Wax, Robert Mundheim Professor of Law at the University of Pennsylvania Law School, graduated with a B.S. from Yale in 1975, and holds an M.D. from Harvard and a J.D. from Columbia. She trained as a neurologist at New York Hospital in the early 1980s, served as a law clerk to Judge Abner J. Mikva on the D.C. Circuit Court of Appeals, and from 1988 to 1994 worked as an attorney in the Office of the Solicitor General at the Department of Justice, where she argued fifteen cases before the United States Supreme Court. She taught at the University of Virginia Law School before joining the faculty at the University of Pennsylvania in 2001. Her areas of teaching and research include civil procedure, remedies, employment law, social welfare law, and the law and economics of work and family. She has written on social welfare issues for the *Wall Street Journal* and is a member of the MacArthur Foundation project on law and neuroscience. Her new book, *Race, Wrongs, and Remedies: Group Justice in the 21st Century*, was published by the Hoover Institution in Spring 2009.

Jeremy Rabkin
Professor of Law
George Mason University
School of Law

Richard J. Zeckhauser
Frank Plumpton Ramsey Professor
of Political Economy
Kennedy School of Government
Harvard University

Research Staff

Gerard Alexander
Visiting Scholar

Ali Alfoneh
Visiting Research Fellow

Joseph Antos
Wilson H. Taylor Scholar in Health
Care and Retirement Policy

Leon Aron
Resident Scholar; Director of
Russian Studies

Michael Auslin
Resident Scholar

Claude Barfield
Resident Scholar

Michael Barone
Resident Fellow

Roger Bate
Legatum Fellow in Global Prosperity

Walter Berns
Resident Scholar

Andrew G. Biggs
Resident Scholar

Edward Blum
Visiting Fellow

Dan Blumenthal
Resident Fellow

John R. Bolton
Senior Fellow

Karlyn Bowman
Senior Fellow

Alex Brill
Research Fellow

John E. Calfee
Resident Scholar

Charles W. Calomiris
Visiting Scholar

Lynne V. Cheney
Senior Fellow

Steven J. Davis
Visiting Scholar

Mauro De Lorenzo
Resident Fellow

Christopher DeMuth
D. C. Searle Senior Fellow

Thomas Donnelly
Resident Fellow

Nicholas Eberstadt
Henry Wendt Scholar in Political
Economy

Jon Entine
Visiting Fellow

John C. Fortier
Research Fellow

David Frum
Resident Fellow

Newt Gingrich
Senior Fellow

Scott Gottlieb, M.D.
Resident Fellow

Kenneth P. Green
Resident Scholar

Michael S. Greve
John G. Searle Scholar

Kevin A. Hassett
Senior Fellow; Director,
Economic Policy Studies

Steven F. Hayward
F. K. Weyerhaeuser Fellow

Robert B. Helms
Resident Scholar

Frederick M. Hess
Resident Scholar; Director,
Education Policy Studies

Ayaan Hirsi Ali
Visiting Fellow

R. Glenn Hubbard
Visiting Scholar

Frederick W. Kagan
Resident Scholar

Leon R. Kass, M.D.
Hertog Fellow

Andrew Kelly
Research Fellow

Irving Kristol
Senior Fellow Emeritus

Desmond Lachman
Resident Fellow

Lee Lane
Resident Fellow; Codirector of the
AEI Geoengineering Project

Adam Lerrick
Visiting Scholar

Philip I. Levy
Resident Scholar

Lawrence B. Lindsey
Visiting Scholar

John H. Makin
Visiting Scholar

Aparna Mathur
Research Fellow

Lawrence M. Mead
Visiting Scholar

Allan H. Meltzer
Visiting Scholar

Thomas P. Miller
Resident Fellow

Hassan Mneimneh
Visiting Fellow

Charles Murray
W. H. Brady Scholar

Roger F. Noriega
Visiting Fellow

Michael Novak
George Frederick Jewett Scholar
in Religion, Philosophy, and
Public Policy

Norman J. Ornstein
Resident Scholar

Richard Perle
Resident Fellow

Tomas J. Philipson
Visiting Scholar

Alex J. Pollock
Resident Fellow

Vincent R. Reinhart
Resident Scholar

Michael Rubin
Resident Scholar

Sally Satel, M.D.
Resident Scholar

Gary J. Schmitt
Resident Scholar; Director of
Advanced Strategic Studies

Mark Schneider
Visiting Scholar

David Schoenbrod
Visiting Scholar

Nick Schulz
DeWitt Wallace Fellow; Editor-in-Chief,
American.com

Roger Scruton
Resident Scholar

Kent Smetters
Visiting Scholar

Christina Hoff Sommers
Resident Scholar; Director,
W. H. Brady Program

Phillip Swagel
Visiting Scholar

Samuel Thernstrom
Resident Fellow; Director, AEI Press;
Codirector of the AEI
Geoengineering Project

Bill Thomas
Visiting Fellow

Alan D. Viard
Resident Scholar

Peter J. Wallison
Arthur F. Burns Fellow in
Financial Policy Studies

David A. Weisbach
Visiting Scholar

Paul Wolfowitz
Visiting Scholar

John Yoo
Visiting Scholar